How Einstein Created Relativity out of Physics and Astronomy

T0180653

Astrophysics And Space Science Library

For further volumes:
http://www.springer.com/series/5664

David R. Topper

How Einstein Created Relativity out of Physics and Astronomy

 Springer

David R. Topper
University of Winnipeg
MB, Canada

ISSN 0067-0057
ISBN 978-1-4899-9493-6   ISBN 978-1-4614-4782-5 (eBook)
DOI 10.1007/978-1-4614-4782-5
Springer New York Heidelberg Dordrecht London

Printed on acid-free paper

Springer is part of Springer Science+Business Media (www.springer.com)

For
Romi
Steve and Hope

Preface, Acknowledgments, and Notes on Format

There are already a zillion books on Einstein and/or relativity. So why did I write this one? There are several reasons. Many good books that explain relativity are out of print. Those still in print often lack a biographical component. There are many very good biographies of Einstein, but the discussions of science are erratic – from poor to adequate, with only a few being quite good. Nonetheless, even the good ones mainly focus only on Einstein's physics, with minimal information on the larger historical context of his science. Mostly they contain only brief discussions, a few sentences or a short paragraph, on say Galileo's or Faraday's or Newton's work that influenced him. There is a critical and vital difference between some physics in the context of Einstein's life and the fuller and deeper milieu within the history of physics – the latter framework being implied in the title to this book. This leads me to a further rationale.

I was first exposed to Einstein's theory of relativity from popular books explaining the theory in simple terms. Later, the theory was taught in my university courses, from which I learnt more through lectures and especially by problem solving. But I only fully *understood* the theory when I studied it historically as a graduate student. The goal of this book is to track the history of the theory of relativity through Einstein's life, with in-depth studies of the background, tracing ideas through earlier scientists. A perusal of the Table of Contents shows that sometimes entire chapters are on this larger context.

So now, there are a zillion-and-one books on Einstein. This one, I hope, is an explanation of the world of relativity, based on an extensive journey into earlier physics and a simultaneous voyage into the mind of Einstein, written for the curious and intelligent reader. If, furthermore, you are holding a paper version in your hands, you may note that another goal was to keep it reasonably short.

* * *

A transparent theme of this book is Einstein's indebtedness to other scientists, despite his self-imposed isolationism. I, as well, do not work in a vacuum. I wish to thank my long-term friend and former student, Wayne Choma, along with my friend and colleague from Physics, Dwight Vincent. I am also especially grateful to my newer friend, Martin Clutton-Brock, Professor Emeritus of the University of Manitoba, who has been a continuous source of inspiration and help in this endeavor. Nonetheless, I alone am responsible for any errors, gross or otherwise, that the diligent reader may find.

My teaching career was almost exclusively at the University of Winnipeg, to which I am especially thankful for a study leave to complete this book. Further appreciation goes to the students in my 3904/3 course on Einstein in the fall of 2011, who allowed me to expose them to a draft of this book, looking for the logic, clarity, coherence, and correctness of my argument; I hope that, in the process, I taught them something about Einstein and relativity.

I cannot end this register of gratitude without acknowledging my wife and companion, Sylvia, who has endured these many years of Einstein in her life. He was seemingly everywhere – on the couch, on the kitchen table, in the bathroom, in the bed, with us on holidays…. She is now quite free of Einstein omnipresent, except for this one book.

* * *

On the format of this book, I have several points to make. The recent decline of real footnotes at the bottom of pages in books baffles me, given the ease and simplicity of inserting them with word processing programs. I understand that editors wish to appeal to the common desire of many readers to have a continuous narrative of flowing words in books. Thus footnotes – if they are used at all – are now usually relegated to the back of books and called endnotes – a sort of residue of marginalizing the unwanted by forcing them to the back of the bus. I am deliberately bucking this trend by belligerently keeping my footnotes where they belong: at the foot of the page. For the life of me I cannot see how these interfere with the flow of the text above: the reader may ignore them by a mere turn of the page (real or virtual), if so desired.

Personally, I recall with pleasure my early university years when I was first introduced to scholarly books and was exposed to some texts that occasionally contained pages where the area of the footnotes was larger than that of the narrative text. I learnt to appreciate the information in these notes, to see how they put the text into a larger scholarly context, and eventually I found myself sometimes reading almost solely footnotes and avoiding much of the text in some books. This experience is in contrast to that of Noel Coward, the late witty writer, composer, and singer, who compared reading footnotes to having to answer the doorbell while in the midst of making love.[1] I prefer to think of footnotes as presaging today's links of hypertext – while I just ignore the doorbell.

* * *

However much footnotes may or may not interfere with the flow of the text, the core narrative of this book, I wish to point out, is intermittently interrupted with

---

[1] Quoted in Grafton, [79], pp. 69–70, and which provides me with my first informing, imparting, irritating, or interrupting footnote – depending on your point of view.

short summaries of overt scientific material. These *Summary* sections (there are six of them) are set apart by being in *italics*.

* * *

The **Bibliography**, placed at the end, contains all material cited in the book. The idiosyncratic citation method used in the footnotes is as follows:

Author(s) or alternative term [item number in the Bibliography] [date of the original document or publication, if relevant], page(s).

Example: Einstein [38] [1923], p. 483.

Decoded: This work, by Einstein, is item number 39 in the Bibliography; it was published in 1923, and I am citing page 483.

I have used extensively *The Collected Papers of Albert Einstein* (Princeton: Princeton University Press, 1987+), a projected 25+ volume series, published in the chronology of Einstein's life (1879–1955); at present it is up to Volume 12 (up to 1921). Each volume consists of a hardbound book of primary sources in the original languages (mainly German) and a paperback book of English translations of selected documents. Each document is given a unique number. I cite this source as:

Einstein Papers, Vol. #, Doc. #.

Since the document number is the same in the original language volume and in the English translation, it is usually unnecessary to specify which book is used. This method simplifies the citation, since obviously the page numbers differ in the two books. Occasionally, however, I do cite specific pages. The symbol **ET** means the English translation.

* * *

Notation in the text of the book for figures is: Fig. 4.2 means Chap. 4, Fig. 2. I have drawn all 30 figures; those few that are copies of images published elsewhere are drawn to avoid copyright infringement. The present obsession over ownership of images, and intimidation of litigation for violations, is a bane to contemporary scholarship. Luckily, the originals of most such images can quickly be found on the Internet.

Finally, I would be remiss if I did not mention what is the ultimate depository of all of Einstein's writings: the Einstein Archives housed in Jerusalem. In his last will, he left his entire estate to the Hebrew University in Jerusalem. Today, the Archives remain on the campus of the University. For information on the Archives, as well as the Einstein Papers Project (mentioned above), go to: http://www.albert-einstein. org/.index.html.

Contents

Part I   Genesis: Riding a Beam of Light

1   Galileo Discovers Inertia & the Relativity of Motion 3

2   Einstein's First Famous Thought Experiment 9
    2.1   Summary 13

3   Einstein: From Zürich to Bern & the <u>Annus Mirabilis</u> 15

4   Converge, Convert, & Conserve: Physics Before Einstein 25
    4.1   Summary 39

Part II   Special Relativity

5   Einstein 1905: The Theory of Relativity Is Born 43
    5.1   Summary 45

6   The Michelson-Morley Muddle 49
    6.1   A Note to the Reader 53

7   What Does the Theory Predict? 55
    7.1   Summary 61

8   From Railroad Time to Space-Time 67

9   1911: The Paradox About Time 77

10   Is the Theory True Today? 79
    10.1   Three Comments for the Advanced Reader 80

Part III   General Relativity

11   Galileo Discovers How Bodies Fall 85

12   1907: Einstein's Second Famous Thought Experiment 89

13 Enter, Mach's Principle; or, Seduced by an Idea 97

14 Einstein's Epic Intellectual Journey: 1907 to 1915 103

15 1916: The Great Summation Paper on General Relativity 113
   15.1 Summary 117

16 1920: Year of Fame, Year of Infamy 121

17 1922: What is Time? Bergson Versus
   Einstein ... and The Prize 127

18 1931: Einstein's First Visit to Caltech 137

19 Is the Theory True Today? 143

Part IV Cosmology

20 Cosmological Conundrums and Discoveries Since Newton 149

21 Einstein 1917: Modern Cosmology Is Born 159
   21.1 Summary 162

22 Three Challenges to Einstein's Cosmic Model 165

23 1931: Caltech, Again; Einstein Meets Hubble 171

24 Cosmology Since 1931: Highlights and Episodes 179

Part V Exodus: Quest for a Unified Field Theory

25 Roots of, and Routes Toward, Unification 195

26 1931: Caltech, Once More 199

27 Exit, Mach; or, the Perils of Positivism 207

28 The Quest...and the Quarrel Over Quanta 215

29 Legacy: From Pariah to Posthumous Prophet 231

Bibliography 235

Index 247

List of Figures

Fig. 2.1   Einstein's thought experiment of riding a beam of light 12

Fig. 3.1   Young's interference experiment. 21

Fig. 4.1   Øersted's experiment 32

Fig. 4.2   Ampère's experiment. 32

Fig. 4.3   Faraday's drawings. 33

Fig. 4.4   Faraday's experiment. 34

Fig. 5.1   Einstein's asymmetry argument 44

Fig. 6.1   The Michelson-Morley experiment. 50

Fig. 7.1   Einstein's thought experiment on the relativity of time 56

Fig. 8.1   The concept of distance in one, two, and three dimensions. 71

Fig. 10.1   Diagram for a simple mathematical derivation
            of the time dilation, using a light-clock. 81

Fig. 11.1   Galileo's thought experiment for falling bodies. 86

Fig. 11.2   Abstract illustration of falling bodies of different weights 88

Fig. 12.1   Einstein's 1907 thought experiment on the identity
            of gravity and acceleration 90

Fig. 14.1   Diagrams showing two unsuccessful ways of applying Einstein's
            equivalence argument 104

Fig. 14.2   Euclid's fifth postulate. 105

Fig. 14.3   Non-Euclidean space. 106

Fig. 14.4   Einstein's prediction of the bending of light by gravity. 108

Fig. 14.5   Le Verrier's discovery of the advance of the perihelion
            of Mercury 110

Fig. 15.1   An illustration on the bending of light by gravity based on the
            equivalence principle. 114

Fig. 17.1   Einstein's resolution of the twin (or clock) paradox 128

Fig. 18.1   Sketch of the group photograph taken after Einstein's
            lecture at Caltech in January 1931 139

Fig. 19.1   Newton's drawing of his thought experiment of a falling
            projectile body going into orbit 145

Fig. 20.1   Stellar parallax. 152

Fig. 20.2   Leavitt's law... 155
Fig. 21.1   Einstein's cosmological model of the universe 161
Fig. 23.1   Hubble and Humason's linear correlation between
             the redshifts and the distances of nebulae 172
Fig. 24.1   The great wall of galaxies.. 185
Fig. 24.2   COBE. The background radiation from the Cosmic
             Background Explorer satellite 186
Fig. 24.3   Variations of the expanding universe... 187

Photos and Credits

Photo 3.1   Einstein and Marcel Grossmann, 1899. Permission:
              Hebrew University of Jerusalem, Albert Einstein
              Archives, courtesy AIP Emilio Segre Visual Archives   16
Photo 3.2   Einstein with wife Mileva and son Hans Albert,
              circa 1904. Reproduced by permission of The Huntington
              Library, San Marino, California   18
Photo 18.1   Einstein at the Mount Wilson Observatory's Hale Library,
              Pasadena, January 1931. Permission: Hebrew University
              of Jerusalem, Albert Einstein Archives, courtesy AIP
              Emilio Segre Visual Archives..   139
Photo 22.1   Einstein at the Leiden Observatory with, Eddington,
              Lorentz, Ehrenfest, and de Sitter, September 1923.
              Permission: AIP Emilio Segre Visual Archives   167
Photo 28.1   Niels Bohr and Einstein, circa 1925-1930. Restoration
              of original negative and print by William R. Whipple.
              Photograph by Paul Ehrenfest, courtesy Emilio Segre
              Visual Archives   222

Part I
Genesis: Riding a Beam of Light

The laws of nature are such that it is impossible to construct a
perpetuum mobile [that is, a perpetual motion machine]....
How, then, could such a universal principle be found [for the
behavior of light]? After ten years of reflection such a principle
resulted from a paradox upon which I had already hit upon at
the age of sixteen....

(Einstein, <u>Autobiography</u>, 1949)[1]

At age sixteen Einstein daydreamed about an imaginary journey. The paradox pon-dered in that dream resurfaced a decade later at the center of what became his theory of relativity.

This teenage fantasy – or what historians of science prosaically call a "thought experiment" – involved an imaginary ride on a beam of light. It went something like this, beginning with a question: How would the world appear if you rode on a beam of light? To reconstruct Einstein's answer we will use the mental image of a traveler (you) in the cockpit of a spacecraft.[2] If you are moving at a speed less than the speed of light, the cockpit will be lit by the light source behind you. But; if you speed-up such that the spacecraft reaches light-speed, the room will turn dark, even though you did not turn off the light source! Why? Because the light source is behind you and the light-beam cannot catch-up with the objects in the room, all of which are also moving at light-speed. This means that when the room becomes dark, you know that your spacecraft is moving at the speed of light. You know this, exactly.

This realization, however, was a problem for the young Einstein, because you are not supposed to know when that exact speed is reached. Why? The reason harkened back to a discovery of Galileo called the principle of the relativity of motion, and

---

[1] Einstein 1979 [1949], p. 49.

[2] Einstein usually used trains as examples, since that was the mode of travel in his time. But space-travel was already a fantasy in literature.

this experience of riding a beam of light contradicted that principle. This is the "paradox" that Einstein "already hit upon at the age of sixteen," as quoted in the epigraph at the start of Part I. To understand his resolution we need to see what Galileo said. Here's that story. (The eager reader may jump to Chapter 2, or even the *Summary* at the end.)

Chapter 1
Galileo Discovers Inertia & the Relativity of Motion

In the 1590s, while teaching at the University of Pisa, Galileo Galilei tried to solve a puzzle. It involved an apparent contradiction between what he was convinced was true about the world, but which our experience of the world implied was not true. The consensus among almost everyone, scientists and others, was that the Earth was not only the center of the universe, and that it remain fixed and unmovable in this place, but that our experience of everything going around us – Sun, Moon, stars, all things up in the sky – was proof that it could not be otherwise. It is impossible for the Earth to move, for if it did, we would experience that motion in various ways. At the very least, clouds and birds could not catch-up with a moving Earth. Or, a weight dropped from a tower would not fall vertically to the bottom, but would fall "behind" the tower. But these things do not happen: therefore, not only is there nothing in our experience of the world that contradicts a stationary Earth, but everything we see and experience actually supports an immovable Earth. So why was Galileo puzzled? Because he was one of the few thinkers at this time who was increasing convinced, obstinately, that the Earth <u>does</u> move.

Such an idea was seriously put forward in mid-century by the Polish astronomer, Nicholas Copernicus.[1] In 1543, shortly before his death, he published his life-work, a hefty book he called "On the Revolutions," the phrase referring to rotations of the heavenly bodies – Sun, Moon, and planets – around a center, which he proposed was really the Sun and not the Earth. This was the first serious challenge, since at least about the third century BCE, proposing a Sun-centered – or, heliocentric – model.[2] To be sure, the common-sense, geocentric (Earth-centered) model was taught in all the schools – Greek, Roman, Christian, Jewish, Muslim – down to Galileo's time. Indeed, Galileo himself taught just such a model in his classes in Pisa. What he was paid to teach from the curriculum, however, he did not necessarily believe. Rather, as early as the 1590s, he was searching for evidence and arguments to support a moving Earth.

---

[1] (1473–1543).

[2] Helios is Greek for Sun. A Greek astronomer, Aristarchus, possibly put forward this idea seriously in the Third Century BCE, but it went nowhere.

D.R. Topper, *How Einstein Created Relativity out of Physics and Astronomy*, Astrophysics and Space Science Library 394, DOI 10.1007/978-1-4614-4782-5_1,
© Springer Science+Business Media New York 2013

One argument that later became the foundation of the principle of the relativity of motion, he presented as an imaginary (or thought) experiment, although clearly based in part on actual experience. Think of yourself in a cabin below deck in a ship, with water dripping from a container hanging from the ceiling, butterflies flitting about, and some fish swimming in a bowl. When the ship is at rest in dock, the water drips vertically to the floor, the butterflies fly anywhere with equal ease, and the fish swim likewise in all directions. Next consider the ship moving at any speed but always at a constant rate; according to Galileo, the experience is the same as when the ship is at rest. There is no preferred direction for the fish or butterflies, nor do they move with different speeds or different degrees of difficultly in specific directions. As well, the water drips vertically downward. The behavior of them all is the same as whenever the ship is at rest. Drawing on personal experience, Galileo wrote: "In confirmation of this I remember having often found myself in my cabin wondering whether the ship was moving or standing still; and sometimes at a whim I have supposed it going one way when its motion was the opposite."[3] Later examples of this human experience would involve trains, airplanes, and spacecraft (such as space-stations).

It is important to realize that this experience is not due to the air moving along in the cabin, for Galileo knew that the same phenomena would occur even if all the air were evacuated (producing a vacuum). All objects in the cabin are moving at the same speed independently of the cabin itself. That's why the water drops do not fall toward the back of the ship, nor do the butterflies cluster toward the back wall. The realization of this behavior of moving bodies contained at its core the important concept of inertia. Although the term itself was neither coined nor used by Galileo, he undoubtedly conceived of it in his thinking about motion, and I will use the word inertia throughout this Chapter on him.[4]

Galileo presented the concept as plainly as any modern textbook, for here are his words: "a heavy body…will maintain itself in that state in which it has once been placed; that is, if placed in a state of rest, it will conserve that [state]; and if placed in [constant, uniform] movement….it will maintain itself in that movement."[5] By putting this phenomenon within in the framework of "states" of rest or "states" of motion Galileo went to the heart of the matter, which shows how radical and innovative his thinking was. Obviously a body in a state of rest resists any change of state, since a push or pull is needed to make it move; this resistance to motion, this tendency to remain static or inert, is the original source of the term inertia from (as seen in a previous footnote) Latin for "inert." Galileo, however, extended – or, more correctly, completed – the concept by including a body in motion, too. The idea that motion is a "state" like rest was his unique conceptualization.

---

[3] Galileo [71] [1632], p. 188.

[4] The German astronomer, Johannes Kepler, introduced the term inertia from Latin, for inert. Kepler was thinking about the need to push a weight that is at rest in order to get it moving, since the weight resists this motion or pushes back against the force making it move. That is, the weight has a tendency to remain inert, or at rest.

[5] Galileo [70] [1613], p. 125.

This was a significant conceptual break with the past. After all, it had always been assumed that for a body to move, something must make it move. The source of motion could be some power within the body, such as a life force in a plant driving it to grow, or the force of a person or animal pushing or pulling a cart; furthermore, to continue this motion required a constant force – stop the force and the plant stops growing or the cart stops moving. From this point of view, rest and motion are polar opposites.

Galileo, however, made the imaginary leap that in the absence of a medium that slows-down a moving body by friction (thus he was thinking of motion within a vacuum), the body will move forever, since once it is put in this state (of motion) there is nothing to take it out of that state, except an external force, and without that stopping-force the body continues to move. The concept of inertia now encompassed a body's resistance to being stopped when in motion, in the same way as a body at rest resists being moved from a state of rest. Whether at rest or in motion, the body resists any change imposed externally: either by making it move from rest, or changing its speed when moving. In both cases it is resisting a change of state. Listen again to Galileo's ground-breaking words: "a heavy body…will maintain itself in that state in which it has once been placed; that is, if placed in a state of rest, it will conserve that [state]; and if placed in [constant, uniform] movement….it will maintain itself in that movement." Rest and motion were not opposites; rest and motion were both equally states of a body – the telltale idea that exposed Galileo's modernity.

* * *

Not withstanding its brilliance, this idea was neither immediately accepted nor even understood. One objection was that a body moving forever required an infinite power to keep it going. But an infinite power cannot be packed into a finite body. Such an objection was conceived within the viewpoint of looking for a cause of motion. Within the framework of causality, the question asked is "Why?" Why does a body move or continue to move? Something must cause this to happen; and if motion goes on forever, the cause must be infinite…so the reasoning went.

Now consider this: a body at rest surely does not require a cause for staying at rest. If one asks why a body at rest does not move, the answer is simply because no one moved it. It is merely an observational fact; a body at rest does not move. But give it a touch. At first, the body does not move – until we apply a force strong enough to make it move. When the body resists being moved – when it, so to speak, pushes back when we push on it – we speak of its tendency to remain at rest or to being inert, and say that this experience of resistance is due to inertia (or say the inertia in the body is resisting the push). At most, inertia is a word (or a concept entailed in a word) describing how a body naturally behaves. It does not require a causal explanation to answer why it does not move when at rest, nor why it resists being moved. These are facts about, or descriptions of, the behavior of bodies. Said another way: this is a shift from a why (causal) to a how (descriptive) explanation of motion.

Now make the next key conceptual step: apply the same argument to a moving body. This step, or leap, or insight was first taken by Galileo, and it completed the concept of inertia by conceiving of a state of motion as being just another state of a body. The body in motion played by the same rules as the body at rest. From this new point of view a moving body (that is, moving at some constant speed) resisted any change of motion (that is, a change of speed) and this opposition to change was said to be due to inertia – the body's tendency to remain in its state of motion. That is how a body moves, and continues to move, possibly forever, if there is no force applied to change its speed. Thus we arrive at the non-causal concept of inertia, and there is no need to posit some infinite internal power. This is just the way moving bodies behave. Galileo therefore introduced an entirely new way of thinking about how things move – to him we justifiably credit the basis of the modern idea of motion. Three hundred years later Einstein, using Galileo's insight, introduced a new way of thinking about time and space, mass and energy, and gravity.

Einstein, in fact, realized how intellectually tortuous was Galileo's route to the concept of inertia. (Indeed, it has taken me over 1,700 words to get to this point in the argument.) In a collaborative book written with a colleague, Einstein recognized that "this law of inertia cannot be directly derived from experiment, but only by speculative thinking consistent with experience." In particular, he emphasized that Galileo's method was based on "thinking of an idealized experiment."[6] This idealization was the abstract world of the vacuum, devoid of friction. Einstein went on to quote a famous line from Newton: "Every body perseveres in its state of being at rest or of moving uniformly straight forward, except insofar as it is compelled to change its state by forces impressed."[7] Could the influence of Galileo be any more transparent?

Today's definition of inertia in any textbook is a direct descendent of Newton's. Inertia is still a non-causal concept of motion. From this point of view there is no answer to the question, "What is inertia?" – beyond its role as a description of how bodies behave when moving or at rest.[8]

This brings us back to the idea of relativity: for the concept of inertia is the keystone of the principle of the relativity of motion, along with Galileo's descriptive experience of things (butterflies, dripping water, and the like) in the cabin of a moving ship. All these bodies are either at rest (when the ship is in dock) or moving with the ship (when at sea) at the same constant speed that the ship is moving. The inertia of these bodies keeps them moving at the same speed as the ship, independently of whether there is air in cabin or not (air is a factor only for living things to survive). Thus the water drips (or any object falls) vertically to the floor, and the butterflies move freely, as do the fish swimming in the bowl. Said succinctly: everything acts and everything appears as if the ship were at rest. Einstein at times presented this idea experientially, in a negative

---

[6] Einstein and Infeld [59] [1938], pp. 7–8. Leopold Infeld (1898–1968), Polish, Ph.D., 1921 in Cracow, was one of Einstein's collaborators at the Institute for Advanced Study in Princeton, where Einstein worked after 1933, as will be seen later. Pais [162], Chap. 29.

[7] This is the most recent translation of the passage (from the third edition of the *Principia*); see Newton [151] [1726], p. 416. The last phrase in the first (1687) and second (1713) editions is (footnote to p. 416), "…insofar as is compelled to change that state by forces impressed."

[8] We will later come across an attempt at an answer when we discuss something called Mach's principle, a concept about which Einstein obsessed for an extended period of his life.

way: as he might put it, there is no experiment a person in the cabin can perform to prove or detect whether the ship is only at rest or moving at a constant speed. Indeed this is the precise experience to which Galileo was referring when he wrote of being in a cabin and "wondering whether the ship was moving or standing still." Furthermore this experience applies to any frame of reference under the same conditions as the cabin of a ship, and hereafter the cabin or any such dwelling will be called a "system." Surely most readers can recall similar experiences of the relativity of motion in a car, train, or airplane. Einstein thus bestowed the expression "inertial system" to such a defined space moving at a constant speed (or at rest), and this abstract term will be used throughout this book.[9] This idea of an inertial system nicely encompasses the idea of the relativity of motion, since there is no way to detect or absolutely know whether the inertial system within which one is dwelling is moving or not; or said another way, there is no method of knowing what is really at rest, you or the outside world.

Another way of expressing this, which initially seems contradictory, is: the laws of motion are the same (specifically, of the same form) in all inertial systems. This constancy must be true since all things act the same way whether the inertial system is at rest or moving with a constant speed. The apparent contradiction arises since, from this viewpoint, there is an absolutist nature to what otherwise is called the principle of the relativity of motion. This, in fact, was another way Einstein favored expressing the inertial experience: he acknowledged this absoluteness by speaking specifically of the invariance (unchanging or permanence) of the laws of motion in all inertial systems; or, further, calling it an invariance principle. Considering this principle mathematically, it means that equations have the same form in different inertial systems. Such a property of an equation I also call covariance, another term used later in this book.

In the end, as will be stressed, the phenomenon experienced within all inertial systems is the same (there is no contradiction) whether it is formulated in terms of relativity or invariance or covariance – it is merely a matter of semantics and formalisms.

---

[9] The term was not unique to him, although relativity theory put it into common usage. He could have come across it in Mach [134] [1883], pp. 292–293, which Einstein read, as will be seen. Mach probably found the term in the work of German physicist, Ludwig Lange, who is often given credit for coining it. See also Fölsing [65], p. 760, note 14.

Chapter 2
Einstein's First Famous Thought Experiment

Having followed this rather arduous conceptual journey to the principle of relativity, which in the end is quite simple and clear, let's go back to the beginning Part I and recall Einstein's imaginary ride on a beam of light at the age of sixteen. Reprising the problem, which we will now speak of in the language of traveling in an inertial system moving at some given speed, we saw that when the system reached light-speed the space in our dwelling became dark and hence we knew our exact speed, absolutely. It should be clear now what the problem was – what Einstein called a paradox – for this knowledge violated the principle of the relativity of motion. The darkened room would be the experiment, so-to-speak, that detected absolute motion.

Before seeing how he resolved this paradox ten years later, we need to look at the historical context, for around this time he reached an important juncture in his life.

* * *

Albert Einstein was born in 1879 in the town of Ulm in southwestern Germany to unobservant Jewish parents, Hermann and Pauline. It was common among middle class German-Jews at this time to identify themselves as more German than Jewish. While Albert was still an infant, they moved to Munich.[1]

There are many myths about Albert's childhood and youth, most of which are not true. He was not autistic, he was not dyslexic, he was not a slow learner, he did not have ADHD, he was not left-handed, he was not a vegetarian, and he did not do poorly in school.[2] In his short autobiography (written in his late 60s) he speaks of himself as "a precocious young man,"[3] and his sister Marie (whom the family called Maja) speaks of his "remarkable power of concentration" where he would "lose himself...completely in a problem."[4] It is true that he did not like school, especially

---

[1] Munich is about 75 miles east of Ulm.

[2] This sentence is not written frivolously: I have seen Einstein seriously used as a poster-boy for these and other causes. For some such speculations, see Neffe [149], pp. 36–37. For a viewpoint closer to mine, see Isaacson [109], p.12, & p.566 n 15.

[3] Einstein [51] [1948], p. 3.

[4] Winteler-Einstein [214] [1924], p.xxii.

D.R. Topper, *How Einstein Created Relativity out of Physics and Astronomy*, Astrophysics and Space Science Library 394, DOI 10.1007/978-1-4614-4782-5_2, © Springer Science+Business Media New York 2013

the rigidity of the Germanic method of teaching; and he did not hide his feelings from his teachers, so much so that at least in one instance a teacher told his parents that his hostility and misbehavior in the classroom was a poor example for the other students. His attitude probably was less about pedagogy and more an expression of a contrarian behavior that would be nearly unwavering over the years.

Another expression of this nonconformity took place during his preteen years when he became extremely religious (from an orthodox stance) and distressed his parent considerably by admonishing them for their anti-religious outlook. But just about the time he would have been Bar Mitzvah, he discovered science (through reading popular science books, he says) and promptly abandoned organized religion, without Bar Mitzvah. He discussed this transition in his autobiography in two ways. First, as an intellectual and emotional transformation, when he came to see the religious worldview as subjective and solipsistic. He spoke of this as being involved with "the merely personal." Science, in contrast, offered liberation from this subjectivity. As he put it: "Beyond the self there was this vast world, which exists independently of us human beings and which stands before us like a great, eternal riddle, at least partially accessible to our inspection and thinking."[5] The objectivity of this independent, other-world freed him from the subjective world of the "personal" self. It also released him from the fetters of religion, which he came to see as mainly composed of lies, since he said he discovered that much written in The Bible was not true.

This, in turn, was coupled with a second transformation or un-conversion, which he spoke of in rather overblown political terms: "…youth is intentionally being deceived by the state through lies," is how he put it, and, he continued: "Mistrust of every kind of authority grew out of this experience."[6] Did he really feel this deeply on political and religious matters around the age of twelve or thirteen? Seemingly he did, because he made a most intriguing political decision a few years later. Here is the background to the story.

Albert's father was a businessman who was either incompetent or unlucky. His business was the electrical industry,[7] which was to the late-nineteenth century what the high-tech industry was to the end of the last century (read: "dot-com"): many tried, few survived. Hermann Einstein was in the latter category. When his business went under, his family was forced to move in with his brother: but Hermann's brother lived in northern Italy. Since Albert was in his last year of Gymnasium (high school), it was decided the family would leave him in Munich in a boarding house while his parents and sister moved to Italy. The arrangement was a mishap: Albert was not able to cope with his loss of his family and went into a deep depression, so much so that he left school without a degree. Before heading for the border, however, he obtained a letter from his mathematics teacher confirming that he had completed the curriculum.

---

[5] Einstein [51] [1949], pp. 4 and 5; Schilpp (ed.) [179], p. 5. The first phrase of this quotation ("Da gab es draussen diese grosse Welt,…") is usually translated as: "Out yonder there was this huge world,…" I believe my translation is closer to what Einstein was expressing.

[6] Einstein [51] [1949], p. 5.

[7] A photograph of the interior of the Einstein electric company and the machinery is in Renn (ed.), [172], Volume One, p. 133.

Contrary to another myth, Einstein did not have difficulties in mathematics. Indeed, his preteen replacement of religious zeal with scientific fervor involved mathematics too. He was given a book on Euclidean geometry, which he devoured, even trying to prove theorems on his own before reading the solutions in the book. In his autobiography he referred to this math textbook as the "holy geometry book"[8] – an extraordinary phrase for such a prosaic subject, but perhaps significantly it was an unconscious reference to his geometry book replacing the previous other "Holy Book" around the same time in his life. He went on to higher mathematical texts, teaching himself and mastering calculus by age sixteen. All of which explains the letter from his mathematics teacher that was in his pocket as he crossed into Italy.

Which brings us back to the political statement: shortly after this episode he renounced his German citizenship, and was therefore living in Italy as essentially a stateless person. This is quite a radical act for a person of his age. One wonders what he found so despicable about Germanic culture that would provoke this final severing of national ties.[9]

We can only wonder too about his parents' response with his arrival in Italy – and present-day parlance does apply – as a high school drop-out. His father had plans for Albert to be an engineer, which he thought would be the paramount profession of the next century. Luckily for Albert (and his father, too), the school Hermann wanted his son to attend, the Swiss Polytechnic Institute in Zürich, did not require a high school diploma but instead required a series of rigorous entrance exams. So Hermann sent the necessary forms and fees, and Albert took the exams. Sadly, he failed several of them: nevertheless, he did so well on the science and mathematics sections that he was given the opportunity to apply again the following year. The Institute director recommended that he spend the year at the Kanton Schule in the town of Aarau, not far from Zürich.[10]

The year Einstein spent at Aarau was of paramount importance in his life. The school had a progressive curriculum, based on using visual sources, hands-on learning in small classes, and strong teacher-student interaction. For the first time in his life he was at ease in school, and he did well, passing with high marks, enough to get into the Institute.

He was likewise happy in his personal life at Aarau. He lodged with a family of one of his teachers, and he became very fond of them. He even had an early romance with one of the daughters.[11] It is also reported that while his fellow students at the Kanton Schule spent their spare time drinking copious quantities of beer, Einstein was drinking from a different source: devouring Immanuel Kant's philosophical treatise, The Critique of Pure Reason.[12] Importantly, it was during

---

[8] "das heilige Geometrie-Büchlein" in German; in Einstein [51] [1949], pp. 8 & 9.

[9] A simple explanation, provided by his sister, is that he was avoiding the military draft. See Winteler-Einstein [214] [1924], p. xxi-xxii. Alternatively, Pyenson [169], p. 51, says: "His decision to renounce German citizenship…can be seen as a reprisal against an entire society that had taken away his family's livelihood."

[10] A Kanton is a Swiss entity similar to a state or providence.

[11] In addition, his sister, Maja, later married one of the sons; and a close friend, who we will meet later, Michele Besso, married another daughter.

[12] Quoted in Miller [143]; p. 181 in the 1981 edition. Actually this was his second read; he first perused Kant's book around the age of thirteen.

Fig. 2.1 Einstein's thought
experiment of riding a beam
of light: my reconstruction.
When the spacecraft
reaches light-speed the
cockpit will become dark

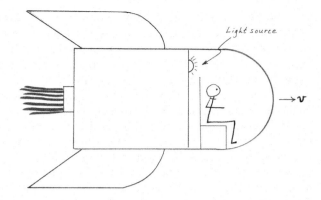

this year that he fantasized about riding a beam of light. Most probably there was something about the ambiance at Aarau that stimulated his creative thinking. Reminiscing about Aarau sixty years later, and only weeks before this death, he wrote: "It made an unforgettable impression on me, thanks to its liberal spirit and the simple earnestness of the teachers who based themselves on no external authority."[13] This open and progressive atmosphere was the context of his now-famous thought experiment.[14]

* * *

Returning to the experiment and the puzzle. Einstein encountered at age sixteen a contradiction between the principle of the relativity of motion, and light disappearing in an inertial system moving at the speed of light. The solution he arrived at ten years later, which (to be seen) appeared in his landmark paper of 1905, is extraordinarily simple. Start with Galileo's idea of inertia, which elegantly explained the motion of bodies, and return to Fig. 2.1: if the motion (or speed) of light is independent of the motion of the system (or the source), then light will fill the room even at the speed of light.[15] This was Einstein's answer: light travels at the speed of light no matter what the speed of the emitting source. The paradox is resolved – but at a price; for it assumes that light behaves differently from other things. Consider throwing a ball with a given force; the ball will travel a specific distance based upon the initial force. If, however, you are running at some speed, the ball will travel further than if you throw it while at rest. In fact, the speed of the ball will be the sum of the speed you throw it from rest and the speed at which you are running. This addition (or subtraction, if you are running backwards) of speeds is a fundamental law of motion.

---

[13] Quoted in Holton [99], pp. 390–391.

[14] For some thought-provoking ideas about Einstein's background and personality see Pyenson [169], Chap. 3, "Einspänner: the Social Roots of Einstein's World View."

[15] A noted before, Fig. 2.1 is my reconstruction of Einstein's thought experiment. There are many attempts at historically retelling Einstein's idea but there is no definitive one because he never gave us one. It appears only in his autobiography, it is very short, almost cryptically written, and with a confusing if not contradictory sentence. I analyze this sentence in some detail in Topper [198], pp. 12–13. Recall too that the autobiography was penned in 1947, over fifty years after the event, which further clouds the historical record.

It should be pointed out that this applies to the case of motion in otherwise empty space, since a medium (air, water, and so forth) slows the speed of balls, other matter, and light too. Also, the phrase "otherwise empty space" is inserted because at the time most scientists held to the model of light as a wave in a diffuse medium known as the aether. (This topic will be explored soon.) As a result, experiments measuring the speed of light at the time were deemed to be measuring the wave-speed within the aether.

So Einstein, in order to resolve the paradox, bestowed upon light a special property that was not true of matter. It became in 1905 a postulate in his theory of relativity; stated simply it asserts that the speed of light is independent of the motion of its source. It follows from this that the speed of light one measures when the source is at rest is the same number no matter how or where it measured. If you are moving toward a light ray the measured speed is the same as when measured at rest; the same applies if you are moving away from a light ray. Always the same number. The specific number, incidentally, was measured with considerable accuracy throughout the nineteenth century, and was near the present value of 186,290 miles/second or 299,790 kilometers/second. Today's symbol little-c (which we will print in bold as **c**) was invented at this time for light-speed,[16] although it was not used consistently until well into the twentieth century.

To recap, Einstein's postulate, said in everyday terms, is this: if I turn on a flashlight while running, the beam of light is emitted at the same speed no matter how fast I go. This is a strange way for something to move – an action that, surely does not apply to my experience with throwing a ball.

The postulate that light acts differently from matter may be viewed from an absolutist viewpoint, too, since the speed of light is the same value in all systems. Behaving so, it took on the nature of being an invariant quantity. This also meant that it had the status of a universal principle, thus fulfilling Einstein's quest mentioned in his autobiography, and quoted in the epigraph beginning Part I. "How, then, could such a universal principle be found [for the behavior of light]? After ten years of reflection such a principle resulted from a paradox upon which I had already hit upon at the age of sixteen...." The principle? – the invariance of light-speed. Furthermore, as a result, there were two invariants in inertial systems – the equations of motion (as seen) and now the speed of light.

* * *

2.1 Summary

At the age of sixteen Einstein envisioned a mental puzzle that involved riding a beam of light. Ten years later he resolved the paradox with two assumptions. The first assumption was the principle of relativity, which asserted the impossibility of an observer in an inertial system knowing absolutely if the system is at rest or in motion. As a result, the equations of motion have the same form (invariant) in all inertial systems.

---

[16] Supposedly the symbol **c** was based on the Latin word <u>celeritas</u>, meaning very fast.

This principle had its origin in the work of Galileo and his attempt to prove a moving Earth, as asserted by Copernicus, despite evidence to the contrary. The concept of inertia was the key idea that led to the principle.

Einstein's second assumption, which he did not borrow from anyone, was the invariance of the speed of light (c). He showed that if the speed of light is independent of the motion of the source, then the paradox is resolved. In an inertial system, the speed of light is always c and everything behaves the same as if the system were at rest, obeying the principle of relativity.

* * *

These two postulates will constitute the only assumptions in Einstein's first publication on relativity. From them he deduced several predictions about the world based upon those postulates. These deductions are the subject of the next Chapter.

Chapter 3
Einstein: From Zürich to Bern & the Annus Mirabilis

After the year spent at the Kanton Schule in Aarau, Einstein was admitted to the Institute in Zürich, and began as a student seeking an engineering degree (as his father wished) but later switched to being a teacher – obtaining in the end a teacher's certificate for physics and mathematics. At the Institute (hereafter the ETH)[1] he spent more time in the laboratories than in the classroom, enjoying the hands-on approach of experimental physics, a behavior probably going back to his childhood time spent in his father's factory.

The idea of Einstein as a nascent experimentalist may come as a surprise, for the conventional image of him is as a theoretical physicist writing equations, typically on the back of an envelope. That indeed was his <u>modus operandi</u> later, but not during these university years. Fortunately he had a very close friend, Marcel Grossmann, who diligently took copious class notes.[2] Einstein being, well, Einstein, was able to pass the tests using Grossmann's notes and reading up on the topics in the textbooks (see Photo 3.1) Two other friends he made from those years were the Italian student, Michele Besso (mentioned above),[3] and a Serbian woman, Mileva Marić. The friendship with Besso was life-long; they corresponded into the last year of there lives, dying about a month apart in 1955.[4]

Mileva was the only woman in the class and their friendship quickly turned romantic. When the relationship became uncomfortably intimate for Mileva, she left the ETH and took classes instead at Heidelberg University.[5] The friendship, however, continued by correspondence, and in one interesting letter Mileva praised a lecture

---

[1] The Institute's name was later changed to the Eidgenössische Technische Hochschule (Federal Technical University); for short, ETH. Sometimes it is called the Swiss Federal Polytechnic, the Polytechnic, the Institute, or just the ETH. Most scholars now use ETH, and I will do so hereafter.

[2] Marcel Grossmann (1878–1936). Of Hungarian-Jewish ancestry, he obtained his Ph.D. in mathematics at the ETH, and became Professor of Mathematics there. He and Einstein would later collaborate on publications (to be seen).

[3] Michele Angelo Besso (1873–1955), of Swiss-Italian-Jewish ancestry, became an engineer.

[4] Besso died on March 15th; Einstein on April 18th.

[5] Heidelberg, in southwestern Germany, is about 150 miles north of Zürich.

D.R. Topper, *How Einstein Created Relativity out of Physics and Astronomy*, Astrophysics and Space Science Library 394, DOI 10.1007/978-1-4614-4782-5_3,

Photo 3.1 Einstein and Marcel Grossmann in the Garden of the Grossmann family house near Zurich, May, 1899. Permission: Hebrew University of Jerusalem, Albert Einstein Archives, courtesy AIP Emilio Segre Visual Archives

by "Prof. Lenard" on the kinetic theory of gases. "Oh, it was really neat," she wrote,[6] and went on to describe how he calculated actual colliding molecules. Albert and Mileva's mutual passions, clearly, were not only of the flesh. Ironically, Philipp Lenard, a later Nobel Prize winner (1905), would enter Einstein's life in a more sinister role (to be seen in Chap. 16). Einstein eventually coaxed Mileva back to Zürich, and they resumed their relationship, with them eventually marrying in 1903.

Einstein graduated in 1900 with respectable marks. It is difficult to assess his marks in terms of present-day systems but they transform to a range from about A- to B-. He hoped after graduation to obtain a position at the school or elsewhere, which was common for the better students. But Einstein, ever the contrarian, was seen by his teachers as an uncooperative and occasionally hostile student, and no one came forward to lend a hand in getting him a job. After graduation, he was on his own. On several occasions he obtained temporary teaching positions. He also found some modest work tutoring

---

[6] *Einstein Papers*, Vol. 1, Doc. 36. In German, "*O das war zu nett…*"

in mathematics and physics. At this time he wrote in a letter that he was "quite exceptionally pleased" with teaching, and in fact was surprised that he enjoyed it as much as he did.[7] On their predicament, Mileva wrote to a close friend that "we have the misfortune that Albert has not found a job," except for some private tutoring, and "it is hardly likely that he will get a more secure position soon; you know my darling has a very wicked tongue and on top of it he is a Jew." She goes on:

> From all this you can see that the two of us make a sorry couple. And yet, when we are together, we are as cheerful as anyone…. And, you know, in spite of all the bad things, I cannot help but love him very much, quite frightfully much, especially when I see that he loves me just as much.[8]

About this time he began working on his Ph.D. at the University of Zürich. His life – or, more correctly, Albert and Mileva's life – became complicated when Mileva became pregnant. This episode in Einstein's life was only known by a handful of very close friends and family until the 1990s, when love letters between them were released. We now know that Mileva went back to her family in Serbia alone to have the child (a girl, named Lieserl), who probably was adopted there. Worth quoting here is a letter Albert wrote to Mileva when she was at her parents' home with the newborn Lieserl, in which he began – prior to asking how the baby was – with physics: "I have just read a marvelous paper by Lenard on the production of cathode rays by ultraviolet light. Under the influence of this beautiful piece of work, I am filled with such happiness and such joy that you absolutely must share in some of it."[9] The joy of physics[10] trumped the joy of the birth of his first child – an omen of dire things to come that should have been a warning to Mileva. In the end Albert never saw his daughter. There does not seem to be any trace of her in Europe, despite extensive sleuthing by numerous "detectives" ever since this story broke into the news.[11]

While this drama was unfolding, Grossmann again came to Einstein's rescue. Through the efforts of Grossmann's father, Einstein was interviewed for a position in a patent office in the Swiss town of Bern. While waiting for the position to materialize, he placed an advertisement in the Bern newspaper for anyone wishing to learn "physics for three francs an hour," and a Jewish-Romanian philosophy student, Maurice Solovine, responded. Thus began another friendship that lasted lifelong, some of which we can reconstruct through the letters that Solovine saved.[12]

---

[7] *Einstein Papers*, Vol 1, Doc. 115, p. 177 ET. Letter of 1901.

[8] Marić [136] [1901], p. 79.

[9] *Einstein Papers*, Vol. I, Doc. 111.

[10] Later, in his 1905 paper on what became the quantum theory of light, Einstein referred to Lenard's experiments on the photoelectric effect as "pioneering work." Einstein (1905), quoted in Stachel [191], p. 193. As noted before, Lenard received the Prize in 1905, interestingly, the year of Einstein's paper.

[11] Zackheim [219]; Isaacson [109], pp. 84–89; Neffe [149], pp. 94–99.

[12] Einstein [54], is a collection of the extant letters: the quotation about the tutoring price is in the introduction by Solovine on page 6. Einstein's last (undated) letter to Solovine refers to the theory of the nonsymmetrical field (he also mentions his "rather serious anemic condition"), thus placing the letter near his death. Solovine lived from 1875 to 1958.

Photo 3.2 Einstein with wife Mileva and son Hans Albert, circa 1904. Permission: Hebrew University of Jerusalem, Albert Einstein Archives, courtesy AIP Emilio Segre Visual Archives

Shortly after meeting Solovine, another student joined-up, Conrad Habicht, who later obtained a PhD in mathematics and became a high school teacher.[13] The three became close intellectual friends, meeting weekly and reading books in philosophy and science. They called themselves the Olympia Academy, which lasted for about two to three years.[14] This brainy pre-Internet chat group was seminal in Einstein's life, for the discussions stimulated his thinking about space and time. He eventually got the job at the patent office, where, delightfully it turned-out, his good friend Besso worked too.[15]

   He and Mileva were married on January 6, 1903. Witnessed by Solovine and Habicht, it was a secular affair: Einstein being a non-practicing Jew and Mileva a Serbian Orthodox Christian. No family member of either spouse was present.[16]

---

[13] Habicht lived from 1876 to 1958.

[14] Hoffmann [97], pp. 37–39; Pais [162], Chap. 29.

[15] Reiser [171], p. 66.

[16] Isaacson [109], p. 85.

Their first son, Hans Albert, was born in May, 1904 (see Photo 3.2). Their second son, and last child, Eduard (also know as Tete or Tedel) was born in July, 1910.[17]

***

Returning to physics, and what historians of science call the miracle or marvelous year, often Latinized as Annus Mirabilis. In 1905, from March to September, Einstein produced the following papers, each a landmark in its field.[18]

- March, the light-quantum paper.
- April, he finished his Ph.D. dissertation[19] and it was published it in 1906.
- May, the paper on Brownian motion.
- June, the first paper on relativity (later called the special theory).
- September, a second relativity paper containing (in a different form) $E = mc^2$.

The papers were published in a major German journal Annalen Der Physik (Annals of Physics).

And so it went, year after year, throughout his sixty-year career. In what is still the best book on Einstein's science, Abraham Pais comments on Einstein in the last years of his life. Referring to Einstein at the time working on proofs of a revised edition of a book on relativity, Pais interjects essentially a rhetorical question: "Does the man never stop"?[20] It seems he seldom did for over six decades from that famous year.

Luckily we can get a glimpse into Einstein's mind in 1905 through two extant letters to Habicht,[21] in which he commented on these papers, one by one. In the first letter, written about mid-May, he asked Habicht for a copy of his thesis and promises "four papers in return." The first (the March paper) on "radiation and the energy properties of light" he said (prophetically) "is very revolutionary." (Little did he know how so!) In the second (the April dissertation) he deduced "the true sizes of atoms." The third (the May paper), explained what "physiologist have observed" as Brownian molecular motion. The fourth (the June paper), which he called "only a rough draft" presented "a modification of the theory of space and time" – a modification, indeed – let alone a rough draft! In the second letter (penned between June and September) he mentioned an idea that "did cross my mind" from the previous (June) paper; specifically an idea that "requires that the mass be a direct measure of the energy contained in a body." So the origin of $E = mc^2$ was an idea that crossed his mind, and a few months later was brought to fruition.

---

[17] Hans Albert died in 1973. Eduard, who was diagnosed later in life with schizophrenia, died in 1965, essentially estranged from his father. The heartbreaking details are in Neffe [149], Chap. 10, especially, pp. 199–205.

[18] All papers are in Stachel [191].

[19] He wrote on the title page: "Dedicated to my friend Dr. Marcel Grossmann."

[20] Pais [162], p. 182.

[21] *Einstein Papers*, Vol. 1, Docs. 27 & 28.

Einstein's last line to Habicht is surely worth quoting in full, although I'm not sure from what frame of mind to read it. He wrote: "The consideration [apparently of the idea that energy is equated to mass] is amusing and seductive; but for all I know, God Almighty might be laughing at the whole matter and might have been leading me around by the nose." One senses that Einstein was aware that he had put forth a breadth of work of potentially considerable significance.

***

The focus in this book is confined mainly to the theory of relativity in Einstein's life. But occasionally his work in other fields must arise, particularly his important contributions to quantum physics and other topics from 1905 into the 1920s, for they have bearing on relativity and his thinking on it. An essential and informative place to commence the topic of Einstein and quantum physics is the provocative thesis of the late and well-known historian of science, Thomas S. Kuhn.[22] He made a strong case that although the German physicist Max Planck[23] was credited in most conventional accounts of the history of quantum physics as initiating the theory – specially with his concept of, and use of the term, "quantum of energy" to explain results of experiments on black-body radiation – his conception of these elements of energy in the original formation of the theory was that they were not discontinuous (that is, not fully quantized). Being, in fact, continuous, they were not actual quanta of energy, despite their name. Furthermore, Kuhn argued, it was Einstein and the physicist Paul Ehrenfest,[24] in papers of 1906–1907 – working independently, and unknown to each other, although they later became very close friends – who really quantized the quantum theory by keeping the quantum of energy fixed, and they, not Planck, should be credited for the actual beginning of quantum physics.[25] There was considerable controversy over Kuhn's thesis initially,[26] but its qualified endorsement has entered the mainstream of secondary sources.[27]

On Einstein's specific 1905 papers, the paper on Brownian motion and his Ph.D. thesis (the latter being initially his most widely cited paper by scientists), gave a

---

[22] To the larger intellectual audience he is better known as a philosopher of science for his book, *The Structure of Scientific Revolutions*, 1962. Interestingly, this work is generally ignored by historians of science, despite it pervading influence on philosophers, sociologists, cultural studies theorists, and many others. The short version of his influence is that he introduced the word "paradigm" to popular culture.

[23] (1858–1947).

[24] Paul Ehrenfest (1880–1933), of Austrian-Jewish ancestry, obtained his Ph.D. from the University of Vienna. See Pais [162], Chap. 29.

[25] Kuhn [126]; the thesis was originally put forward in 1978.

[26] For example: Klein et al. [119], an essay review of Kuhn's book in <u>Isis</u>. As well, there is much discussion in these essays comparing Kuhn's historiographical approach in this book with that in *The Structure of Scientific Revolutions* (1962), noted in a previous footnote. See also, Topper [196].

[27] Brush [20]. In Brush's review he refers to the essay by Olivier Darrigol [32], who agrees with Kuhn. Also Brush [19], pp. 121–123, and [16], p. 92.

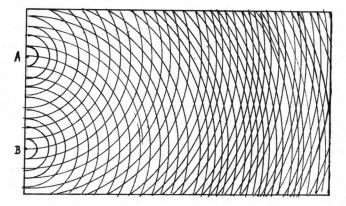

Fig. 3.1 Young's interference experiment. Young's geometrical drawing shows how two light beams produce light and dark bands. To see this, observe the image from an angle, horizontally. Initially, he used this diagram to illustrate water waves interacting

major boost to the atomic theory. Surprising as it may seem, in the early years of the last century there still was no consensus on the actual existence of molecules or atoms. But as evidence mounted, the atomic nature of matter was eventually accepted, but not until the second decade of the century. A key part of that story was the atomic model of electrons circling a nucleus put forward in 1911 by the New Zealand-born British chemist and physicist Ernest Rutherford, and the resulting rise of a new branch of physics.[28]

The light-quantum paper broke with the nineteenth century wave model of light and reintroduced a discrete theory of light into physics. Isaac Newton[29] had put forward a particle theory in the seventeenth century based on his work on the spectrum of light, and his authority dominated light theory thereafter. Newton darkened a room and through a pinhole in the window he projected an image of the Sun through a prism and onto a screen, obtaining an oblong spectrum of colors from the red to the violet.[30] Thomas Young in the early nineteenth century discovered the interference of light, a phenomenon that could only be explained on a wave model. Young sent a beam of light simultaneously through two holes (Fig. 3.1), and projected the light onto a screen, resulting in the production of alternating bands of light and dark on that screen. (Hold Fig. 3.1 at a sharp angle and you will see beams of light and dark emanating from the two sources. It is quite astonishing that this

---

[28] Rutherford (1871–1937) is given major credit for the discovery of the atom's nucleus, which gave birth to nuclear physics. Incidentally, the discovery of the electron will come up later (Chap. 7).

[29] (1642–1727).

[30] There was also a wave model of light that originated in the seventeenth century, and which continued to be explored by some scientists throughout the next century, but the dominant model was Newton's particle model, even though he too had flirted with a wave model. For this and remainder of this Chapter, see Topper [197], pp. 124–129, & 132–133.

mere geometrical diagram can reproduce the phenomenon.) How could two beams of light-particles add together to make dark bands? By an analogy with water waves, which were known to interact together and produce higher and lower waves, and even to cancel out.[31]

Young modified he experiment using different colored light and found at the two extremes, red and violet, the bands were spread the farthest apart and closest together, respectively. This meant that on the wave model colors were a function of their wave-lengths, with red having the longer and violet the shorter wave-lengths. Not long after Young introduced the wave model – which was vehemently debated, since Newton's particle model was well-entrenched in the scientific community – the wave model was found to be adaptable to a further modification based on another discovery. A measurement of the temperature of the light spectrum of the Sun showed that different colors had different temperatures, with the red side being, seemingly, the hottest – but, surprisingly, moving the thermometer beyond the red gave an even hotter temperature. This meant a part of the spectrum that was not visible actually existed. This discovery was easily adapted to the wave model – just add a longer wavelength beyond the red – although it was strange to discover something invisible. It was known that there was such a thing as heat without light, namely radiant heat (think of a radiator), and experiments showed that radiant heat behaved the same as light. This invisible light with the longer wavelength beyond the red, in time, became known as infrared light.

No sooner was invisible light found beyond the red part of the spectrum, than the budding field of photography opened-up the other end of the spectrum, with the shorter wavelengths. Photographs of the light spectrum found a darkened area beyond the violet. Hence, ultraviolet light was discovered.[32] By the mid-nineteenth century the wave model replaced Newton's particle theory of light, mainly because it was the only model that worked to explain the interference phenomenon, with further evidence coming from the invisible ends of the spectrum. In the late-nineteenth century the spectrum was even expanded further with x-rays found beyond the ultraviolet, and microwaves beyond the infrared.[33] Light therefore was more than meets the eye, with most of what constitutes the light spectrum being invisible to the human eye.

---

[31] Young's discovery is found in a series of lectures he delivered to the Royal Institution. It should be pointed-out that Young's diagram was initially used by him to show how water waves interact. It appears in lecture XXIII on hydraulics. See Young [218], I, p. 290 & Plate XX, Figure 267, p. 777. Then in lecture XXXIX, on light and colours, when he discusses the interference of light, p. 464, he refers to the same diagram, and now makes the analogy between light waves and water waves. Rothman [175], pp. 12–21, goes so far as to speculate that Young may not have actually performed these experiments with light, but came to his conclusions through the geometry of the wave model.

[32] The two terms were not agreed upon until late in the century. The prefixes infra- (below) and ultra- (beyond) were often used for both colors.

[33] The next century would add "radio" waves (AM, FM, TV, and radar) beyond the infrared and gamma rays beyond x-rays.

Accompanying the wave model was necessarily the concept of a medium to carry the wave, the best candidate being the aether[34] – an idea of a medium filling space that went back to the ancient Greeks. Since the late Middle Ages it was known that sound is a wave in air, and in the seventeenth century it was shown that no sound exists in a vacuum; sound needs the medium of air to produce waves. Therefore light required a medium to produce light waves; how could it be otherwise? In the next Chapter we will see how the properties of this aether were bound up not only with the nature of light but with the new science of electromagnetism.

***

That story will be saved for later. For now let's briefly return to the relativity papers of 1905, and Einstein's relationship with Mileva. Along with the revelation in the 1990s of the birth of Lieserl, another exposé supposedly appeared in the love-letters – that Mileva was really a co-discoverer of the theory of relativity. The sources of this conjecture were several phrases in their correspondence where Albert spoke of "our paper," and "our work," or "our research." The popular press at the time picked up on this, with a number of articles and essays claiming that Mileva deserves credit for Einstein's theory (implying, as well, misogynous factors); there was even a PBS documentary implying as much. It is true that Albert, despite his otherwise radical notions, had a very nineteenth century masculine mindset with a less than liberating attitude toward woman, a fact that undoubtedly emerged later in their relationship, as their marriage broke down. The misogyny charge held, but not Mileva's supposed claim to a piece of the theory; the latter has not withstood the test of significant subsequent historical efforts.[35] In the end, the overwhelming evidence showed that the relativity theory (vis-à-vis Mileva) was Albert's alone – independently of various love-speak rhetoric in their letters.

---

[34] I prefer using the original spelling of aether to distinguish it from the modern compound, ether.

[35] A good summary of the controversy is Martinez [137]. For what it may be worth: over her lifetime, Mileva never made any claim to having played a role in the theory.

Chapter 4
Converge, Convert, & Conserve: Physics Before Einstein

Before returning to the theory of relativity – although the restless reader may skip right to the next Chapter, and Part II; or just go to the summary on p. 39 – we must probe more deeply into the physics background, especially specific topics from the seventeenth to nineteenth centuries, in order to fully grasp the context of Einstein's ideas. This historical survey will constitute the remainder of this Chapter and set the stage for the next.[1]

A good place to begin is with the actual title of the 1905 paper on what became relativity theory. This may come as a surprise to some, for it is: "On the Electrodynamics of Moving Bodies."[2] Electrodynamics? That is, electricity and magnetism. What do they have to do with the relativity of light and motion? Isn't relativity likewise more about clocks and time? Clearly there is a story here. Indeed there is, and in many ways it is one of the most fascinating stories – although not told often enough – in the history of science. We start near the beginning.

Probably our first knowledge of electricity was what we call today static electricity, where rubbing the fur of an animal causes the fur to stand up or to attract powders or other hairs in the vicinity. A closer look at static electricity led to what the ancient Greeks saw as action-at-a-distance; where things attracted or repelled each other without there being a visible intermediary. These hidden powers later were referred to as "occult" (from Latin for "hidden"), a much less cumbersome term than action-at-a-distance. In addition to the Greeks' knowledge of static electricity they knew of the power of the lodestone, a type of stone that attracted small metallic objects. Lodestones were natural magnets and they too had the occult powers of invisible attraction and repulsion. These two cases of occult powers discovered in the ancient world commenced a long history – from the Greeks, through the Romans, and throughout the Middle Ages. The word occult certainly has further connotations, bordering on spiritualism, psychic powers and such, which are related historically to the topic here. Well into the nineteenth century, various masters of

---

[1] Much of the history in this Chapter may be found in Topper [197], pp. 132–136.

[2] Or, in the original German: *"Zur Elektrodynamik bewegter Körper."*

D.R. Topper, *How Einstein Created Relativity out of Physics and Astronomy*, Astrophysics and Space Science Library 394, DOI 10.1007/978-1-4614-4782-5_4,
© Springer Science+Business Media New York 2013

the occult practiced this "black art," such as Franz Mesmer (of "Mesmerism"), who used electrical and magnetic contraptions to ply their wares – but that is another story.[3] Suffice to point out that what is relevant to this narration is what led to Einstein's thinking in 1905.

Aristotle, who put forward the basic theory of a physics of motion that was still taught by Galileo around 1600, had serious misgivings about action-at-a-distance, arguing that the idea was an unscientific form of superstition. He asserted that motion was possible only by either internal powers in nature (such as a plant growing from a seed), or from an external push or pull by direct contact. Therefore static electricity and the power of a lodestone were caused by two types of rotating media, more diffuse than air, which are otherwise invisible. The authority of Aristotle over two millennia made occult powers questionable components of legitimate science.

The history of occult powers is coupled with astrology, which was another "art" vying for legitimate status over the ages, since astrology was based upon the necessity of invisible powers operating between heavenly bodies (Sun, Moon, stars, and planets) and earthly events. The phenomenon of tides, for example, which were realized in the Middle Ages to be correlated to the periods of the Moon and Sun, was interpreted by some as evidence for both occult powers and astrology. Further support for occult powers was supplied by a discovery and invention transported to the West from China – the magnetic compass. Here was an object that always pointed in one direction, north–south. Beyond its obviously critical use as a navigational instrument, it was a scientific puzzle: what hidden power was moving this magnetic needle to line-up only one way? An answer followed from the question: What is the north–south direction in a geocentric cosmos? Answer: it is the celestial world of the stars, and from the viewpoint of those living the northern hemisphere, the stars were rotating around the north-star (called Polaris, the pole star, by the Middle Ages); this star was obviously the object to which the compass was "pointing." This discovery supported further the association of astrology and occult powers.

The Scientific Revolution of the sixteenth and seventeenth centuries was a transitional period in this story, not only with the shift from a geo- to a helio-centric cosmos and the corresponding discredit of astrology as a pseudoscience, but specifically for the empirical study of magnetism and later electricity through the work of the Englishman William Gilbert[4] and his landmark book De Magnete (1601). The most famous discovery in his extensive work was the magnetism of the Earth itself, a realization that explained the alignment of the compass as a having a terrestrial cause, since the Earth itself is a magnet. Further work on magnetism and then electricity followed, with the next century seeing extensive progress in both fields. Some highlights were the construction of large static electricity machines from which electricity was able to be "conducted" along wires, and then the invention of the battery in 1800 that launched electrical science into the nineteenth century as a fundamental branch of physics.

---

[3] See, e.g., Darnton [31].
[4] (1544–1603).

Prior to a deeper look into why Einstein's 1905 paper had that seemingly odd title, I need to pick up another thread from the seventeenth century. For the demise of occultism was only partial, and this was because of Isaac Newton. Newton's Principia (1687), as is often correctly said, was a culmination of the Scientific Revolution (some scholars would say the culmination), with Newton's laws being at the core of a physics of motion that lasted until Einstein. When the book was published, however, one key dispute involved the very topic of occultism. It is true that Newton was able to give a mathematical explanation of gravity to explain objects falling on the Earth, and even the possibility of an artificial satellite launched around the Earth; and he used the same mathematics to explain the motion of the Moon around the Earth, and even the planets and comets around the Sun. The mathematical rule was an inverse-square law; that means there was an attractive power between any two bodies of matter, and this power diminished as the distance-squared. If two bodies move apart twice their original distance, the force between them diminishes by one-fourth, or triple the distance and the decrease is one-ninth, and so forth. As well, this power or force (**F**) was also proportional to the product of the masses of the bodies, say **m** and **M**. Written as a sentence it is: Force (**F**) is proportional to mass (**m**) times mass (**M**) divided by the distance (**D**)-squared. Written as a proportion (where α is the symbol of proportionality) the law is:

$$\text{Force } \alpha \text{ (mass } \mathbf{m} \times \text{ mass } \mathbf{M})/\mathbf{Distance}^2$$
$$\text{or}$$
$$\mathbf{F} \, \alpha \, (\mathbf{m} \times \mathbf{M})/\mathbf{D}$$

This is the mathematical expression of Newton's law of gravity.

It is worthwhile to dwell a bit on the concept of mass, which was unique to Newton, especially because of its relevance to Einstein's theory of relativity. Prior to Newton, weight was the fundamental concept involving heavy bodies. But Newton realized that his theory of gravity meant that the weight of a body was a function of its distance from the center for the Earth. Today a fact often taught to young students is that they weigh different amounts on the Moon and other planets, due to the different masses of these bodies. This bothered Newton because he also knew that, despite this relativity of weight, there was another parameter of a body (call it bulk?) that did not change. He chose the English word, "mass" to express this unchanging concept. The word mass had its origin in the late Middle Ages, and was used usually for a lump of dough or clay. Newton gave it its first scientific meaning. For mainly theological reasons, it was important to him to ground his theory upon absolute (not relative) entities.

The same thing bothered him about time, which also was a relative entity if Copernicus was right. Time would be measured differently on different planets; days, years, and so forth differ depending where you are. Local time was relative. So he searched for an absolute way of measuring time, but did not find it. He relied, intuitively it seems, on what today we call the biological clock; although time may be measured differently on different planets, just as one's mass does not change (only one's weight), one does not age at different local rates. It is no different than the fact that we have on Earth both solar and lunar calendars, which measure different years, yet we all age at the same rate. The absolute time that measures the rate of aging is the clock of the universe – namely, time (absolute time) in the mind of God (as I read Newton). As mentioned before, there was a theological component to Newton's absolutes. In sum, he found that – or set-up definitions such that – mass and time were absolute entities (despite the relativity of weight and local time).[5]

But what about the cause of the inverse-square law of gravity; where does this force come from? The theory worked as a deductive-mathematical system, but what was the source of this power of attraction between all masses of matter? Newton did not say in 1687; instead, he hid it behind the title of his book: the full title being, in English, Mathematical Principles of Natural Philosophy. With the adjective "mathematical" modifying the noun "principles" Newton was confining himself to only a mathematical explanation; or said otherwise, he was avoiding a confrontation with the physical cause of his physics of motion.

Newton's personal struggle with these matters, and his subsequent published and unpublished work on this important topic, are beyond and irrelevant to the story here. What is relevant, however, is that the majority of seventeenth century scientists viewed occultism as pseudoscience or superstition, believing it was their duty to purge real science of this medieval blight. The motion of matter could only come about by direct contact, and this was at the center of what they called the mechanical philosophy. With the machine as their model (think of gears and wheels moving together), any force between, say, the Moon and Earth must take place by an intermediary, say an aether. Otherwise, a power operating across empty space would be action-at-a-distance or occult. Moreover, the power seemed to be instantaneous; if there were no intermediary mechanism, then the Moon, so to speak, was aware of the Sun instantaneously at all times. How could this be? How could gravity travel across space from the Sun to Saturn (the most distant known planet at the time) without a direct physical mechanism? This is the conceptual background to Newton's Principia and explains, as a start, the title – with Newton hiding behind the adjective "mathematical."

Because of this conceptual problem, as Newton's inverse-square law of gravity was adopted as a law of physics, accompanying it was an apparent residual occultism. In essence there was an underlying tension in Newtonian physics as it developed throughout the eighteenth and into the nineteenth centuries. Some scientists thought gravity would ultimately be explained by some sort of aether model. Others devised a philosophy of science based upon a mathematical worldview, where a mathematical explanation was necessarily and sufficiently the essence of science. To support their viewpoint they borrowed a line of reasoning from Newton himself. Near the end of the later editions of his Principia,[6] in a famous passage defending the absence of a physical explanation for gravity, he concluded this way: "And it is enough that gravity

---

[5]It is interesting that Newton coined the word mass to distinguish absolute weight from relative weight, but he did not coin a corresponding term for absolute time: duration? Also, the format of writing equations as sentences first, I borrow from Taylor and Wheeler [195], a excellent, although rather advanced, book on Relativity.

[6]As noted above, the book went through three editions during his life. The quoted passage was added to the second edition (1713) and remained in the third (1726), which was published shortly before his death (1727).

really exists and acts according to the laws that we have set forth and is sufficient to explain all the motions of the heavenly bodies and of our sea [i.e., the tides]."[7] The crux was this: Don't ask further, or deeper, physical questions; the mathematics works, and that is sufficient. Coming from Newton at the end of his treatise, this was seemingly the final word. One way some scientists interpreted this passage was simply to accept the gravitational force as a physical fact and ignore its similarity to occult powers. By the nineteenth century, there was a common view that occultism was astrology, black magic, and other nonsense that too many gullible people blindly believed; in the other hand, science was, well…science, the real laws of nature. Never mind, or just ignore, the apparent occultism underlying the very law of gravity.

Empirical support for Newton's theory came in the late-eighteenth century from the last published experimental work of the English scientist Henry Cavendish. Using a torsion balance, a very sensitive apparatus devised by a colleague, John Mitchell, Cavendish measured the attraction of a large lead sphere on a small one, and confirmed the inverse-square law. This inspired the Frenchman Charles-Augustin Coulomb,[8] who made an analogous series of experiments with a torsion balance to measure electric and magnetic forces. He discovered what became known as Coulomb's laws. He found that the forces between both electrical charges and magnetic poles likewise obeyed an inverse-square law analogous to Newton's law of gravity. Not only was there an inverse-square relation for the distances between the charges of electricity and the poles of magnetism, but the forces were also proportional to the product of the charges and the poles, analogous to the masses of bodies for gravitational force.[9] There was an obvious formal similarity or symmetry among all three laws – gravity, electricity, and magnetism.

* * *

Coulomb's impetus for performing these experiments was the analogy with the inverse-square law of gravity of Newton, but interestingly, having found the same formal force relations, he concluded that the electricity and magnetism (as well as gravity) had nothing in common physically; only the mathematical formulae were the same. His reasons were these: electric poles exist whereas magnetic poles only come in pairs (cut a magnet in half, and you get another magnet), and gravity is only an attractive force, whereas electricity and magnetism are both attractive and repulsive. To Coulomb, the differences among them was greater that their similarities. This also meant that any intimation of a unification of gravity with electricity and magnetism was conceptually meaningless.

A contemporary, who also pondered Newtonian forces, was the German, Immanuel Kant. Better known today as an academic philosopher,[10] in his time he was also a science teacher who read Newton carefully, looking for the philosophical

[7]Newton [151] [1713 & 1726], p. 943.

[8](1736–1806).

[9]Baigrie [5], pp. 43–46.

[10]Recall that Einstein (Chap. 2) read his *Critique of Pure Reason* as a teen.

underpinnings within the <u>Principia</u>. He found three: space, force, and matter.[11] Space was empty, Euclidean (geometrical) space; and gravitational force extended across empty space instantaneously between matter – whether between apples and Earth, Earth and Moon, or Sun and Earth and other planets. Writing a century after the publication of the <u>Principia</u>, Kant was placid about action-at-a-distance; forces filling space presented no philosophical challenge to him. Rather, it was his starting point for a reformulation of Newton's trinity. Space and forces were taken for granted as real. But what is matter? As an academic philosopher, Kant famously wrote that all our knowledge begins with experience.[12] So consider the experience of a stone. It takes up space, a specific volume of space; move the stone and it empties the space where it was, and fills a new space. A stone, thus, is just something that fills a volume of space. How do we know this something? Hold the stone, cradle the stone, now squeeze the stone; it pushes back. I cannot push beyond its surface; I cannot penetrate the surface with my hands. I come to know (experience) the stone by its filling a volume of space, and by its pushing back as I squeeze that volume. The stone is ultimately experienced as a volume of repulsive forces confined to a volume of space. But the stone cannot be that alone; for if it were, this volume of repulsive forces would explode. There must be internal attractive forces too, forces holding back the repulsive ones; these two forces (of attraction and repulsion) are in equilibrium, and this makes up our experience of the stone in space. Moreover, any internal attractive forces that penetrate beyond the stone's surface would then be its gravitational force. Kant thus reduced matter to force. The conception, in turn, reduced Newton's trinity of matter, space, and force to Kant's duality – a universe composed of space and force alone.

Kant's idea was seen by some as a third way, a duality (space and force) between Newton's trinity (matter, force, and space) and alternately the aether model. More importantly, this third way resulted in two key conceptualizations that grew out of Kant's formulation: the unity (or convergence) of forces and the transformation (or convertibility) of forces. Both led to what became known as field theory, one of the unique ideas of nineteenth century physics, and a framework that had a most profound impact on Einstein's conceptualization of physics. The best way to explain this is through case studies of three important scientists.

Kant's ideas had an acute influence on the Danish scientist, Hans Christian Øersted. He surmised that if Kant was right, then there should be a relationship, an interconnection, between all the forces of nature. Of most immediate concern to him were those of electricity and magnetism. Here were two powers in nature, having separate histories, but both being part of the story of hidden (occult) forces and the focus of attention of two branches of physics especially over the last two centuries.

---

[11] His primary writing on this topic is his 1786 book, *Metaphysical Foundations of Natural Science*, a book relatively unknown to academic philosophers but studied today by historians of science; more importantly, it was widely read by some key scientists in the nineteenth century. See Kant [113].

[12] "That all our knowledge begins with experience there can be no doubt." This is the first sentence his *Critique*.

Could there be a connection between them? If so, then it should be made visible by, say, a conducting wire (of electricity) influencing a magnet. In principle, this seemed to be a simple experiment: place a wire with a current on a table and move a compass toward it to see if the needle is deflected. If so, then there would be a connection between electric and magnetic forces.

Ørsted performed this experiment in 1820. The year before, in a well-respected encyclopedia, the case was made that magnetism and electricity were two entirely independent forces in nature, as Coulomb asserted, based on the physical differences between them. Ørsted, however, took his cue from Kant's speculation on the unity of forces and tried the experiment. Historians unfortunately cannot reconstruct precisely how his discovery was made, for there are several different accounts of what happened. It seems it went something like this: at first he placed the conducting wire on the table pointing east–west and moved the compass (pointing north–south) toward to wire; the compass, being perpendicular to the wire, would, he expected, rotate parallel to the wire. But nothing happened. He was disappointed, but still sure there was a connection. He was so sure, in fact, that he later showed the experiment to a group of students to demonstrate what he had done, even though the experiment had been a failure. Apparently, in this instance, the compass was on the table and Ørsted place the live wire across the compass, parallel to the needle, either by choice or by accident. Well, the needle moved, rotating perpendicular to the wire! Success: Ørsted discovered a connection between electric and magnetic forces; the two actually interacted with each other.

But what a strange connection it was. The needle pointed perpendicular to the wire, not parallel, as expected. Of course, this explained why the first experiment failed, since he set up the perpendicular arrangement at the start, and hence the compass did not move. But why did the compass move perpendicular to the wire? It was as if gravity attracted an apple by making it fall across, not toward, the Earth; or if you pushed a rock forward and it moved sideways. Clearly these forces were different from gravity, which acts along a straight line between two masses.

The news of this discovery traveled quickly among scientists and it was easily reproduced, since most scientific labs had this simple equipment available. A closer look at the phenomenon, specifically from a 3-D viewpoint, revealed the source of the perpendicular force (Figure 4.1). The magnet's perpendicular interaction with the wire formed a circle around the wire, or a series of circles along the wire. Also, since the magnet had an internal direction (with two poles, N-S), that direction was found to be dependent on the current in the wire; that is, reverse the current, and the arrow in Figure 4.1 is reversed. (The arrangement was later called the right-hand rule, found in any textbook today: wrap your right hand around the wire according to the magnetic direction, and your thumb points in the direction of the current.) This discovery led to an important invention by the French physicist, André-Marie Ampère. He made a coil of wire, sent a current through it, and found that it behaved just as a magnet does (Figure 4.2). By reversing the current, the magnetic polarity was reversed. This was the first electromagnet, one of the major inventions of the century.

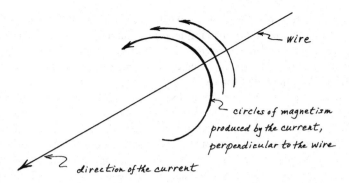

Fig. 4.1 Øersted's experiment. Circles of magnetism are produced by an electric current in a wire

Fig. 4.2 Ampère's experiment. The first electromagnet

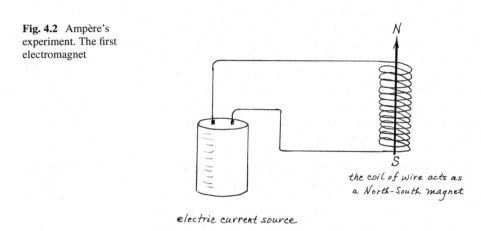

One way of interpreting this invention, as well as the physical interaction between the electricity in the wire and the magnetic needle of the compass, is this: when electricity was sent thought the wire, circular magnetic forces were produced around the wire. Since magnets interact with other magnets, than the circular magnetic forces around the wire caused the magnetic compass needle to line up in circles. But the concept of circular forces was so strange that many scientists did not accept the idea. Even Ampère, who used the phenomenon to invent the electromagnet, made a concerted mathematical effort to find ways of reducing these seemingly circular forces to linear forces (like Newton's gravity), acting-at-a-distance (as implied in Coulomb's Laws).

The next important scientist in this story, Michael Faraday in England,[13] was perfectly comfortable with circular forces because he, as Øersted, was enamored by

---

[13](1791–1867).

Fig. 4.3 Faraday's drawings. Patterns produced when metal shavings are sprinkled on a piece of paper covering a magnet (*top*) and electric charges (*bottom*)

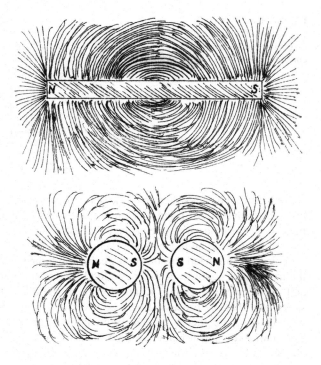

Kant's idea of the unity of forces. Of humble beginnings and little formal education (mainly self-educated), and especially lacking the advanced mathematical knowledge of Ampère, Faraday was forced to rely on his physical intuition and visual imagination. He pictured a wire with a current as in Figure 4.2 surrounded with circles of magnetism (he called them lines of force). The same was true for both magnets (and electromagnets), which he imagined as surrounded by circles of magnetism; and electrical charges had radiating lines of electric force. In a stroke of brilliance, he moved beyond this mental imagery toward actually seeing these lines of force by concocting something that is now done in virtually every elementary science class on the planet: place a piece of paper over a magnet and sprinkle some metal shaving (such as iron filings) on the paper. He did the same for electric charges. Patterns were formed – beautiful, symmetrical patterns, which he drew as in Figure 4.3, where the top figure is a magnet and the bottom two electric charges. Later he introduced the term field (for the first time in 1845, for the magnetic lines of force[14]), and the term subsequently became used for such patterns of lines of both electric and magnetic forces.[15] This was the birth of what became field theory.

---

[14]Faraday chose a common Old English word having Germanic origins, giving it its first scientific usage.

[15]Harman [87], p. 72.

Fig. 4.4 Faraday's
experiment. The moving
magnet produces an electric
current in the wire

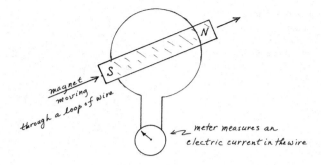

To Einstein the construction of the field concept was "of great importance" in the
history of physics, for it led to a "new reality." As he wrote: "The electromagnetic
field is, for the modern physicist, as real as the chair on which he sits."[16] Faraday too
viewed the field as an independent reality.

This conception led Faraday to the conjecture that Øersted's experiment involved
the actual production of a magnetic field around the wire by the electric current;
and, furthermore, decoding Kant in terms of the convertibility of forces, Faraday
speculated that the process should be reversible so as to produce an electric field
with a moving magnet. It was an argument based on symmetry in nature: if moving
electricity produces a magnetic field, then moving magnetism should produce an
electric field. In principle, although not in practice, the experiment he performed
went something like Figure 4.4. (The actual experiment was more complex.) By
analogy with Øersted's experiment, as the magnet moves, an electric field should be
produced around the magnet, and therefore when it moves through a loop of wire, a
current should arise in the wire and be detected by a meter. In 1831 Faraday succeeded, producing electricity from magnetism, and thus converting one into the
other. As long as the magnet, or a series of magnets, moved through the wire, electricity was produced. Just as Øersted's experiment led to Ampère's invention of the
electromagnet, Faraday's experiment had two spin-offs, the electric generator (or
dynamo) and the electric motor; in the former, motion produces electricity, and in
the latter, electricity produces motion.[17]

Kant's inspiration thus came to fruition with these two famous experiments.
Faraday, in his later years, tried to take his discovery further, by bring gravity into
the conversion process. He wrote in his diary in 1849, "Surely this force [of grav-
ity] must be capable of an experimental relation to Electricity, Magnetism, and the
other forces, so as to bind it up with them in reciprocal action and equivalent
effect."[18] He went on to perform numerous experiments moving electrical devices

---

[16]Einstein and Infeld [59] [1938], p. 151.

[17]These were two devices that Einstein's father and uncle were selling in their electrical business,
and that Albert learned about when he visited their factory.

[18]Quoted in Williams [213], p. 466.

through the gravitational field (note that the field concept was now applied to gravity, too); he expected that an electric current would result, just the way he produced electricity with a moving magnet. But no current was detected. The last paper he wrote for publication was titled, "Note on the possible relation of Gravity with Electricity or heat," but it was rejected by the prospective journal.[19] Einstein spent most of the last 30 years of his life trying, theoretically, to unify gravity and electromagnetism, too. He called this – which is the main topic of Part **V** – his quest for a unified field theory.

Since Newton had turned the physics of motion into a mathematical subject, and hence mathematical physics developed and flourished throughout the next two centuries, it was inevitable that these developments in electromagnetism would need to come within the framework of mathematical manipulations, skills beyond Faraday's ability. The details of the mathematical development of electromagnetism from Faraday to fruition are part of a long and complex story, but one key person essential to it is the Scottish physicist, James Clerk Maxwell.

Using Faraday's field model, Maxwell wrote a trilogy of papers and a book that transformed the physics of electromagnetism into mathematical expressions. Developed over seventeen years, the final formulation required over twenty calculus-based equations. After Maxwell's untimely death from cancer in his late forties[20] a new mathematical notation, vector calculus, was invented. Calculus, in part, is a mathematical system for dealing with changing functions (such as bodies that accelerate, or waves moving); vectors have direction, and are often drawn as arrows. Hence vector-calculus can express changing functions (such as electromagnetism) that also have direction. This notation reduced Maxwell's cumbersome formulation to just four equations: essentially, one for each of Coulomb's laws, and one each for Øersted's and Faraday's laws. Today these are called Maxwell's equations and they form the core of what we call, borrowing Faraday's term, an electromagnetic field theory.

How did Maxwell interpret his mathematical work on Faraday's field? Although he adopted and helped extend the use of Faraday's term, Maxwell did not necessarily concur with Faraday on the independent reality of the field. Rather, he initially viewed the field as grounded in the aether. As stressed in Chap. 3, by mid-century the wave model of light had replaced Newton's particle model, and there was widespread belief that an aether pervaded the universe. This ubiquitous medium, to Maxwell, was the substratum of his equations. What Faraday viewed as lines of force, Maxwell saw as stresses and strains in the aether. What is more, this idea led Maxwell to the next unification of the century, although he did not realize its significance.

Imagine an electric charge with lines of force emanating from it as stresses and strains in the aether. Further imagine the charge vibrating, thus producing wave motions among the lines through the aether. Doing essentially the same mathematically by manipulating his equations, Maxwell deduced an equation expressing a wave, as expected; and, in addition, a closer look at the equation showed that the

---

[19] Williams [213], pp. 465–479.

[20] (1831–1879). Note: he died the same year Einstein was born.

wave's speed was around the speed of light. Did this mean that there was some connection between light and electromagnetism? Faraday himself had speculated on a possible connection between not only electricity and gravity but also electricity and light. Think of a spark; does it not occur as electricity jumps between a wire and a contact? Maxwell never went any further than pointing out the similarity between the motion of light through the aether and the motion of electromagnetic waves. Sadly, he died before realizing the full significance of his deduction.

It was the German physicist, Heinrich Hertz,[21] who experimentally verified what he believed was implied in Maxwell's work: light is actually an electromagnetic wave. In a series of experiments performed in the late 1880s he showed that electromagnetic waves have the properties of light (using invisible light beyond the infrared region). This was another unification for the century: light and electromagnetism. What we call light is merely the visible portion (visible to the human eye, that is) of the entire spectrum of, now, electromagnetic waves.

This unification, to be sure, was a triumph for Maxwell's work. But what was its physical meaning? For many scientists it was simply further support for the reality of the aether, the substratum of light waves which, in turn, were now understood as electromagnetic waves. Hertz too was aware of Maxwell's view of the aether in his field theory, but Hertz also knew of Maxwell's ambivalence on this, especially in his late work where he tried to deduce his equations from a few physical assumptions about the mechanical properties of the aether. Was Maxwell ultimately putting forward a view similar to Newton's later editions of the <u>Principia</u> – that a mathematical theory was a sufficient explanation of reality? After grappling with Maxwell's writing on this, Hertz, perhaps in exasperation, famously wrote: "Maxwell's theory is Maxwell's system of equations."[22] In so doing, Hertz bequeathed to the late-nineteenth century the mathematical world-view of electromagnetism – a fourth interpretation of reality, although one that hearkened back to the title of Newton's <u>Principia</u> and the last sentence of the book .

But there was more: in Maxwell's late work there was another mathematical deduction that he made from his physical assumptions about the aether: he calculated the energy contained within the electromagnetic field. This then bring us to the last unifying concept of the nineteenth century: energy. This indubitably is a concept required, if only for later understanding Einstein's famous equation, $E = mc^2$. As seen, this term was used before in this book – in discussion around the 1905 papers and so forth – for it is common today in our everyday speech. Nonetheless, surprising as it may seem, the word was not used within science until the nineteenth century. It was not used by either Galileo or Newton, although occasionally historians anachronistically, and erroneously, apply the term to them. Energy is an English word, coming into use in the seventeenth century for vigor of expression or action, but never in a scientific context. It was first used two centuries later by Thomas Young

---

[21] Tragically, Hertz had a short life (1857–1894).

[22] Hertz [91] [1893], p. 21.

for a mathematical term that often appeared in problems involving the motion (or velocity = v) of bodies: it was mass (m) times speed- or velocity-squared; that is, mv^2. Later it become $\frac{1}{2}mv^2$, and was called kinetic energy, as it is today.

The full development of the concept of energy, however, took place primarily within the study of heat, not mechanics. Modern thermodynamics grew to maturity during the nineteenth century, having its origins in the technology of the Industrial Revolution. The steam engine, for example, stimulated the study of heat and motion. (Interestingly, this was the causal opposite of electromagnetism, where the science led to the technological inventions.) The eighteenth century concept of heat was primarily based on a fluid model, heat being made-up of an invisible fluid (rather similar to the aether) called caloric (from which our word calorie derives) whose flow, say from a hot body in contact with a cold body, explained why the two bodies come to the same temperature. We still speak of heat flow, even if only metaphorically.

By the early nineteenth century there were several arguments against the caloric model, one being that friction produces heat (think of rubbing your hands to keep warm), which seemed to be able to produce an inexhaustible supply of heat. Besides, where did this heat come from? It apparently came from the mere motion of the hands itself. Now, consider a steam engine: heat produces steam which in turn moves pistons; here is the opposite, heat producing motion (or, as it was called at the time, mechanical work). What we have in these cases is a transformation or a conversion process between heat and motion (or work). Not surprisingly, these processes were compared with the other transformations being discovery at this time, such as between magnetism and electricity. Likewise, and importantly, they were connected: for example, an electric wire can get hot; this then is the case of electricity producing heat, radiant heat (which, as noted before, is the basis of an electric heater or stove).

This study of heat and motion added a very important idea to thermodynamics – the concept of conservation, which was eventually extended into the other phenomena. Due to the nature of heat and mechanical work, both were quantifiable, and hence when the transformation occurred the "before" and "after" amounts could be measured. Assuming a closed system, the conversion of heat to work (or work to heat) was seen as a conservation process; in an ideal system, where no heat was lost during the process, the total quantity of heat was transformed into an equal quantity of motion or work. Similarly one could quantify electricity transformed into the heat, and so forth. There were a number of scientists (theorists and experimentalists) working around these areas of transformation and conservation. In a classic article on energy conservation, Thomas S. Kuhn identified a dozen scientists in the early- to mid-nineteenth century who contributed to its conceptualization.[23] A culmination of this endeavor was a seminal article published in 1847 by the German polymath Hermann von Helmholtz,[24] in which he drew together and summarized the transformations between heat and work, heat and electricity, electricity and magnetism, and even electricity and chemical processes. To be sure, electricity produced by a battery

---

[23] Kuhn [125], pp. 66–104.
[24] (1821–1894).

is an example of a chemical process producing electricity, and thus the link was made between chemistry and electricity. Also, growing out of his work on animal physiology, he added the transformation of food (heat coming from calories) and living movement (or bodily chemistry).[25]

These transformations, coming initially from thermodynamics, were another expression of convertibility, with the added concept of conservation. But what was the something that was being conserved during the conversion? Recall that the work of Øersted and Faraday was within the Kantian framework of the convertibility of forces; Helmholtz too drew upon Kant for inspiration. Thus, the title of his classic paper was "Über die Erhaltung der Kraft" ("On the Conservation of Kraft"); the translation for kraft at the time was "force," betraying the Kantian overtones. Today, a German dictionary translates kraft as power, strength, force, or energy, the last being the scientific meaning. Even among English scientists, there was an evolution of terminology from (Kantian) forces into energy. The first use of the word energy for heat phenomena (after Young's original use for mechanical work) was in 1848/1849 by William Thomson (later Lord Kelvin), within the context of heat being conducted through a solid, when he wrote that "no energy can be destroyed."[26]

Looking at this from a conceptual point of view, we saw that Ampère would not accept the reality of circular forces, trying instead to reduce them to linear action-at-a-distance. Faraday's conception of the field as an independent reality was an alternative view, which did not involve an instantaneous transmission of action between bodies. His was a different conception of force entailing, as well, the idea of convertibility or transformation. With the parallel development of heat theory, and the underlying conceptions of convertibility and conservation, it was almost inevitable that all this would be put together under some conceptual framework with a new terminology. In the end, after an evolution from "force" in English and kraft in German, by the late-nineteenth century, energy (Energie, in German) was generally agreed upon as the word expressing this quantity that was conserved across various transformations in nature. So we find in those latter years the concept of energy being increasingly employed, as the term was disseminated through textbooks. By Einstein's youth, the law of the conservation of energy was a law of nature, the last of the unifications (or convergences) of this remarkable century in physics. Much of this can be summed-up as the three "cons": convergence, conversion, and conservation.

Even so, despite all this convergence and convergence coming to fruition, there was another lingering and deeper question in late-nineteenth century physics, as the young Einstein studied his textbooks. Energy is the "energy" of what? For Faraday, it would be energy in the field; for Maxwell (with some late doubts), energy within the aether. And so forth. Some scientists, naturally were content to understand these processes mathematically (as Hertz's interpretation of Maxwell's equation), and that was enough. Perhaps the most interesting answer to the question is that energy is an independent entity and is itself the ultimate reality; this

---

[25] Harman [87], pp. 41–45.

[26] Quoted in Harman [87], p. 51.

viewpoint was called energeticism. One of the more impassioned energeticists was the physical-chemist Wilhelm Ostwald.[27] Read these excerpts from an 1895 essay called "Emancipation from Scientific Materialism,"[28] the year of Einstein's thought experiment that began this book:

> [E]nergy, it had been urged, is only something thought of, an abstraction, while matter is a reality; [but] exactly the reverse, I reply [is true]. Matter is a thing of thought....[T]he Actual, *i.e.*, that which acts upon us, is energy alone,...and the result is undoubtedly that the predicate of reality can be affirmed of Energy only.[29]

From Kant's reduction of matter to force, and the evolution of force into energy, to Ostwald's further conceptualization of the essence of reality being energy alone, it seems a mere short step to the equation $E = mc^2$ – simply, the transformation between energy and mass. Actually, Ostwald wrote, a few paragraphs following the above quotation: "Matter is therefore nothing but a group of various forms of energy co-ordained in space, and all that we try to say of matter is really said of these energies."[30] Energeticism, therefore, provided another conceptualization or world-view of reality for the late-nineteenth century to ponder – and, in the context of this book, for our young Einstein to ruminate over.

Having mentioned Ostwald in the context of energeticism, let me add a story of cheerless and bitter irony. After Einstein graduated from the ETH, and was desperate for employment, he wrote to Ostwald asking for a job in his lab; he enclosed a copy of his first paper (published in 1901), on capillarity, in which he had cited Ostwald's work. Einstein's father too, wrote to Ostwald, mentioning that Albert felt he was a burden to his family by being unemployed. Neither of the Einsteins ever heard from Ostwald, but we know he received Albert's letter because the copy of Einstein's publication was found among Ostwald's papers.[31] I suspect he received (and ignored) Hermann Einstein's letter, too. There a coda to this story. In 1909 Ostwald won the Noble Prize for chemistry; in 1910 he was the first person ever to nominate Einstein for the Prize in physics.

* * *

4.1 Summary

The title of Einstein's first paper on relativity was, "On the Electrodynamics of Moving Bodies." To grasp the background to the theory, we traced briefly the history of magnetism and electricity from the Greeks specifically to the nineteenth

---

[27](1853–1932).

[28]The phrase "scientific materialism" betrays an ideological (philosophical, even political) context for this paper, which would take us far beyond the scientific matters of this book.

[29]Quoted in Nye [153], pp. 348–349. Emphasis by Ostwald. Inserts are mine.

[30]Quoted in Nye [153], p. 350.

[31]Pais [162], p. 45 & 506; and *Einstein Papers*, Vol.1, Doc. 92.

century; there the fusion of the two took place (one transforming into the other), plus their merger (unification) with light, since visible light is an electromagnetic wave of specific wave-lengths that the human eye can see.

Additionally, the concept of energy grew out of the theory of heat in the nineteenth century, culminating in the law of the conservation of energy, entailing at once unification (convergence), transformation (conversion), and of course conservation.

* * *

Part I began with a quotation from Einstein's autobiography and an analysis of the thought experiment mentioned therein as a resolution. We end by returning to the autobiography and another well-known quotation.

The passage to be quoted followed a discussion of the concept of wonder, which he defined as "an experience [that] comes into conflict with a world of concepts already sufficiently fixed within us." Einstein then recalled:

> A wonder of this kind I experienced as a child of 4 or 5 years when my father showed me a compass. That this needle behaved in such a determined way did not at all fit into the kind of occurrences that could find a place in the unconscious world of concepts.... I can still remember – or at least believe I can remember – that this experience made a deep and lasting impression upon me. Something deeply hidden had to be behind things.[32]

This reminiscence of the incident with the compass (recalled in 1947)[33] is probably true and significant in his memory, as he also mentioned it in the interviews with Alexander Moszkowski in 1919–1920, where the recalled his age as being 5 years old; and again he repeated it to Rudolf Kayser, around 1930, where he dated it as 4 years old.[34] The incident may be viewed within the context of both electromagnetism in his own work – which will play a key role in the 1905 paper through the quest to unify it with gravity – and around rise and fall of his father's fortune in the electrical industry. Looking at it from the framework of Part I, Einstein was exposed as young child to the prototypical case of occult powers in nature – and apparently reacted in, what he called, a wondering way. Finally, how appropriate and striking it is that around the time of his writing his autobiography, as we shall see later in another context, he also spoke of the phenomenon of action-at-a-distance – essentially occult powers – as being "spooky."[35] Such "an experience [that] comes into conflict with a world of concepts already sufficiently fixed within us" can, we may agree, be succinctly expressed as being just plain spooky.

---

[32] Einstein [51] [1949], p. 9. He went on to say that a "second wonder" occurred at the age of twelve, and this was where he mentioned the "holy geometry book," discussed before.

[33] Stachel [192], p. 284 n18, dates the manuscript for the autobiography as written in 1947, so I use that date throughout this book.

[34] Moszkowski [147] [1921], p. 221; and Reiser [Kayser], [171], p. 25.

[35] *Spukhafte*, from a letter to Max Born in 1947, in Einstein [56], p. 155. The context is his assertion "that physics should represent a reality in time and space, free from spooky actions at a distance."

Part II
Special Relativity

...I must confess that at the beginning, when special relativity began to germinate in me, I was visited by all sorts of nervous conflicts.

(Einstein, 1919/20)[1]

In the history of physics there are two years that are so significant that they are identified by the Latin phrase *Annus Mirabilis*. Part of which is explored in more detail here, is 1905 in Einstein's life.

The first, 1666, is when Newton was living on his mother's farm because Cambridge University was closed due to a plague spreading from London.[2] During that time Newton develop his theory of light, his theory of gravity along with the deduction of the inverse-square law, and the formulation of what later was called calculus – developments that changes the course of physics. Born on Christmas Day in 1642 (the same year Galileo died), Newton was in his early- to mid-twenties during his *Annus Mirabilis*.

Einstein, as noted before, was born the same year (1879) in which Maxwell died; and he was, like Newton, also in his mid-twenties during his *Annus Mirabilis*. As we saw, he published a series of papers that changed the course of physics ever since. These papers were on a theory of atoms and molecules, a quantum theory of both light and matter, and the first part of what became known as the theory of relativity.

---

[1] Einstein 1919/20, quoted in Moszkowski [147], p.4.

[2] The sojourn actually lasted about 18 months: from August 1665 to April 1667, with a brief return to Cambridge from late-March 1666 to June 1666. By historical convention, the singular *Annus Mirabilis* is used.

Chapter 5
Einstein 1905: The Theory of Relativity Is Born

The title of this first relativity paper, "On the Electrodynamics of Moving Bodies," revealed that the topic grew out of issues pertaining to electromagnetism, which as seen in the last Chapter was a major field of nineteenth century physics. The first sentence of the paper set-up the conceptual framework: "It is well known that Maxwell's electrodynamics – as usually understood at present – when applied to moving bodies, leads to asymmetries that do not seem to be inherent in the phenomena."[1] What were these supposedly well-known asymmetries, and what were the phenomena he was speaking of? The next series of sentences gave the answers. Einstein went back to Faraday's experiment of a magnet moving through a conductor (or a loop of wire). With the conductor at rest, the moving magnet generated an electric field and this produced a current in the wire (Fig. 4.4). If, on the other hand, the magnet were set at rest and the conductor moved over the magnet, even though (according to theory) there was no electric field around a stationery magnet, a current of the same strength as the former case was still produced in the wire (Fig. 5.1). Einstein saw this as an asymmetry in interpretation in the two cases; one with, and one without, an electric field. Despite this apparent conceptual problem, the "observable phenomenon here depends only on the relative motion of conductor and magnet…," since in both cases electricity was produced in the wire. Undoubtedly, the asymmetry was not inherent in the phenomenon, as he said. From the viewpoint of the phenomenon, a current was produced by any relative motion between the magnet and the conductor, whereas the theory of electromagnetism affirmed that an electric field was produced only by a moving magnet, not one at rest.

At the beginning of this first sentence Einstein also said that this asymmetry was well-known. Is this true? As far as I know, no one pondered this problem until Einstein; the so-called asymmetry was, therefore, "well-known" only to him.[2] Significantly he was very much vexed over this essentially aesthetic problem. In 1920, reminiscing on the genesis of relativity, he stated that Faraday's experiment

---

[1] Einstein (1905), quoted in Stachel [191], p. 123.

[2] Holton [98], p. 7, makes the same assertion.

D.R. Topper, *How Einstein Created Relativity out of Physics and Astronomy*, Astrophysics and Space Science Library 394, DOI 10.1007/978-1-4614-4782-5_5,
© Springer Science+Business Media New York 2013

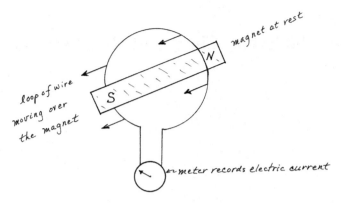

Fig. 5.1 Einstein's asymmetry argument. The reversal of Faraday's experiment (Fig. 4.4). The magnet is at rest and the wire moves over the magnet, and an electric current is again produced

"played a leading role for me when I established the theory of special relativity." He then pointed to the difference in the "two cases" of the magnet at rest and in motion: "The idea that these two cases should essentially be [conceptually or theoretically] different was unbearable to me." Unbearable is surely a strong word; but it led him to conclude that "the existence of the electric field was a relative one" and this, in turn, "forced me to postulate the principle of (special) relativity."[3]

So in 1905 Einstein deduced the relativity of motion principle from his interpretation of Faraday's experiment, in this context, coming from electrodynamics. We saw in Chap. 1 that Galileo deducted a relativity principle from the mechanics of motion. Putting the two together Einstein made this assertion: "Examples of this sort…lead to the conjecture that not only the phenomena of mechanics but also those of electrodynamics have no properties that correspond to the concept of absolute rest."[4] (The phrase deleted from this sentence will be discuss later; it is a topic unto itself.) Here was a generalization of the relativity principle that encompassed both mechanics and electromagnetism. It was, in a partial sense, the postulation of a unification of the physics of motion with electromagnetism.

Looking at the previously quoted sentence again, notice that Einstein used the plural "examples" whereas, in fact, he only mentioned Faraday's experiment. This is a bit of a puzzle. If by examples he is only thinking of empirical evidence, then there is a problem, since he only mentioned one. If he included conceptual arguments (such as thought-experiments) perhaps he had in mind riding a beam of light.

---

[3]"Fundamental Ideas and Methods of the Theory of Relativity, Presented in Their Development," *Einstein Papers*, Vol. 7, Doc. 31, p.135 ET. This unpublished paper (a long 35 page manuscript) from the Pierpont Morgan Library, New York, has been dated as January 1920, based on a series of letters with the editor of the journal <u>Nature</u> that appears to refer to this document. *Einstein Papers*, Vol. 7, Doc. 31, note 3, p. 279. It is probably a draft of an article that was never published in *Nature*: Stachel [192], pp. 262–263.
[4]Einstein (1905), quoted in Stachel [191], p. 124.

Maybe he was thinking of the mechanical examples of Galileo. There was also (as will be seen in the next chapter) an argument against absolutism by the physicist/ philosopher Ernst Mach, who deeply influenced him. Or was a mere slip of the pen? In any case, Einstein was so sure of this "conjecture" that he immediately wrote: "We shall raise this conjecture (whose content will hereafter be called 'the principle of relativity') to the status of a postulate...." This set-out the logic of the paper: start with postulates (which presumably are true) and deduce relations from them. It was the same deductive framework of Euclid's geometry – the "holy book" he read as an early teenager. Euclid's geometry indeed has been a model for logical reasoning; after all, it was the framework of Newton's <u>Principia</u>.

Picking up with the last sentence again, and reading further, Einstein continued: "We shall raise this conjecture...to the status of a postulate and shall also introduce another postulate, which is only seemingly incompatible with it, namely that light always propagates in empty space with a definite velocity $\mathbf{V}$ [later replaced with $\mathbf{c}^5$] that is independent of the state of motion of the emitting body." Now we indubitably know where this second postulate came from: namely, the thought experiment he pondered since the age of sixteen. The conclusion he arrived at with the resolution of the paradox, we now see, became the second postulate of the theory of relativity. Said otherwise: it was the invariance of light-speed.

Notice too his aside that this invariance may appear "incompatible" with the relativity principle. He is probably pointing to the apparent difference between the relativism of the first postulate and the absolutism of the second. We saw above, however, there was no real contradiction, as Einstein openly asserted. Whether one speaks in terms of relativism or invariance, as we know, most of this was semantics only. The two postulates were compatible.

* * *

5.1   Summary

The logical framework of special relativity is based on the model of geometry: start with postulates and deduce results; in this case, physical results.

   It begins with two postulates:

1. *The principle of relativity (applied to both mechanics and electromagnetism)*
2. *The invariance of the speed of light (in a vacuum).*

   The relativity principle asserts that it is impossible for an observer in an inertial system to determine absolute rest or absolute motion for the system. The invariance of light asserts that the speed of light is independent of the motion of the source;

---

[5] As noted before, the symbol little-$\mathbf{c}$ was invented for the speed of light in the nineteenth century, but it still was not commonly used at the time of Einstein's early papers; he used capital-$\mathbf{V}$ for the speed of light in 1905. Nonetheless, hereafter the little-$\mathbf{c}$ notation will be used in this book.

therefore, in empty space light always travels at the same speed c. From these two postulates alone, Einstein derived special relativity.

* * *

Having set-up these postulates, he next affirmed: "These two postulates suffice for the attainment of a simple and consistent electrodynamics of moving bodies based on Maxwell's theory for bodies at rest." There were no more conjectures (or postulates), two suffice; what is more, they produce (which remained to be seen) a "simple" and "consistent" theory. Note, as well, the aesthetic terminology – simple, consistent (along with the previous symmetry/asymmetry) – in Einstein's reasoning. Not withstanding the important role of experiments (Faraday's, for example) and empirical knowledge, such aesthetic matters increasingly loomed large in Einstein's works over the years.[6]

The very next sentence proclaimed the first deduction from the two postulates: "The introduction of a 'light aether' will prove to be superfluous, inasmuch as the view to be developed here will not require a 'space at absolute rest'...." With that clause, Einstein dismissed the aether, bringing back empty space into physics. There are several routes in his scientific life towards this deduction. Chronologically, the first is probably a revealing letter to Mileva dated August 1899. In it he wrote:

> I am more and more convinced that the electrodynamics of moving bodies, as presented today, is not correct, and that it should be possible to present it is a simpler way. The introduction of the term "aether" into the theories of electricity led to the notion of a medium of whose motion one can speak without being able, I believe, to associate a physical meaning with this statement. I think that the electric forces can be directly defined only for empty space...[7]

Parts of this letter were clearly echoed in the first relativity paper six years later. Note too that he had no aversion to empty space, presumably because he, like Faraday, believed in the reality of the field.

Another source of his rejection of the aether was the quantum concept of light he put forth in his first 1905 paper, since it did not require a continuous medium and implied empty space. Besides, the relativity principle itself implied the denial of absolute rest, and this, in turn, negated an aether since such a medium would be the system at rest from which absolute speed could be measured.

All of these were conceptual arguments: but there was an empirical route, as well. A series of experiments performed in the eighteenth and nineteenth centuries attempted unsuccessfully to measure the earth's motion through the aether. This topic brings us back to the phrase that was deleted from the previous sentence that began

---

[6] Shelton [184], makes the case for observation playing a strong role in Einstein's thought; in addition, he makes a point of distinguishing between Einstein's simplicity criterion and aesthetic matters, which Shelton says are usually erroneously seen as identical – and which I confess I am guilty of, since I subsume simplicity under the umbrella of aesthetics.

[7] *Einstein Papers*, Vol. 1, Doc. 52, p. 131 ET.

with "Examples of this sort," and which referred to the phenomenon of Faraday's experiment being due only to the relative motion of the magnet and conductor. Here now is the entire sentence: "Examples of this sort, together with the unsuccessful attempts to detect a motion of the earth relative to the 'light medium,' lead to the conjecture that not only the phenomena of mechanics but also those of electrodynamics have no properties that correspond to the concept of absolute rest." Einstein no doubt was acknowledging empirical evidence against the existence of an aether, and doing so as further support for his relativity principle gleaned from Faraday's experiment. But what were these "unsuccessful attempts to detect a motion of the earth relative to the light medium"? Unfortunately, he did not say. As a result, assumptions and speculations on what experiments he had in mind abound in the historical literature; specifically there is still a major historiographical debate on this that began over forty years ago. The next Chapter tells that story, which it is worthwhile to confront before continuing with our analysis of the first relativity paper. As before, no one is stopping the impatient reader from passing directly to Chapter 7.

Chapter 6
The Michelson-Morley Muddle

The theory of relativity became known to the wider public in the early 1920s, and as a spin-off Einstein assumed the social role of a scientific celebrity. His being photogenic and willing to clown for the cameras helped reinforce his status. The result was a plethora of popular books endeavoring to explain relativity theory to an eager public. This practice continues today – witness the very book you are reading now. Interestingly, one of the first popular books was written by Einstein himself.[1] Published first in German in 1917, it was written almost immediately after he finished the general theory in late 1915. That Einstein felt the need to explain his seemingly impenetrable theory to the general public is of more than passing interest. Apparently he wanted to beat others to the task, and put the watered-down version in his own words.

In the book he stressed the role of what is today is recognized as a famous experiment performed "to detect a motion of the earth relative to the light medium," this being the Michelson-Morley experiment (done in the 1880s). Devised by the American physicist, Albert A. Michelson, with help from a chemist friend Edward W. Morley,[2] the essence of the experiment is this: on a horizontal apparatus that is able to rotate, simultaneously send two beams of light in directions 90° apart, toward mirrors set at identical distances from the source (Fig. 6.1). If the Earth is moving through the aether, with the aether at rest in the universe (or filling Euclidean space), then there should be a constant aether wind blowing towards us in the direction we are moving through it. For the experiment, therefore, there should be a specific angle of rotation of the apparatus, where one beam is parallel to the aether wind and

---

[1] Einstein [49] [1917].

[2] Michelson was on the Physics faculty of Case School of Applied Sciences, later called Case Institute of Technology, in Cleveland, Ohio. Morley was a Chemist at Western Reserve University, the two campuses being adjacent to each other. In 1967 the two amalgamated into Case-Western Reserve University as it is today. I was a graduate student at the time they were joined, and my Master's degree in Physics is from Case Tech, while my Ph.D. in History of Science is from Case-Western.

D.R. Topper, *How Einstein Created Relativity out of Physics and Astronomy*, Astrophysics and Space Science Library 394, DOI 10.1007/978-1-4614-4782-5_6,
© Springer Science+Business Media New York 2013

Fig. 6.1 The Michelson-Morley experiment. Two perpendicular light beams starting simultaneously at **O** reflect off mirrors at equal distances from **O**. At a specific rotation of the apparatus, beam **A** will be parallel to the postulated aether wind due to the motion of the Earth

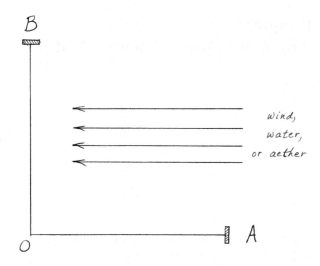

the other perpendicular to it as in Fig. 6.1. Think of replacing the light beams with two airplanes starting simultaneously at the same point (**O**) and flying at the same speed 90° apart and making a round trip of the same distances. If there is a wind blowing, as in Fig. 6.1, then plane **A** slows down on the first part of the trip flying into the wind and speeds up on the way back. Plane **B** travels at a constant speed but with a side-wind slowing it down both to and from the turning point. Michelson pictured boats on a river but in all situations the mathematical deduction is the same: vehicle **B** arrives slightly before **A**. For light beams, likewise, there should be a time lag. The experiment, however, detected no time lag in the beams of light; they always arrived back simultaneously. Michelson was disappointed; he was sure there was an aether (to explain, at least, the interference of light) and was convinced there must be a way of detecting it. Today he is famous for this experiment, but he believed (to his dying day in 1931) that the endeavor was actually a failure.

Today Michelson's experiment is evidence that there is no aether. As noted, Einstein mentioned this in his popular book, and subsequently writers of popular accounts echoed this; but they often carried the logic further, asserting that the experiment was part of the genesis of Einstein's theory of relativity. The negative result supposedly led him to doubt the existence of the aether and supported his argument toward the relativity principle. Variations of this historical reconstruction are found in virtually all popular accounts and introductory textbooks on relativity. In the textbook for a course on mechanics that I took many years ago (a book that is still in print), the chapter on the special theory of relativity starts: "The history of the theory of relativity begins with a famous experiment, performed by Michelson and Morley in 1887, to determine the velocity of the earth in space ...."[3] But a careful reading of Einstein's popular book reveals that he made no such historical argument. The Michelson-Morley experiment is not mentioned in the early section on

---

[3] Fowles [66], p. 257.

the development of the principle of relativity; only later, after the deductions and as empirical support for the theory, is the experiment brought in.[4] It is striking too that Einstein made no mention of this experiment in his autobiography, where, among other things, is found the thought experiment about riding a beam of light.[5]

There is more: we have recently come to know through new documents in the Einstein Archives that as a student at the ETH in Zürich he proposed to his physics professor an experiment similar the Michelson's to test the motion of the Earth through the aether, but the idea was rejected.[6] At the time neither he nor his professor apparently was aware of Michelson's work. Thus, we need to ask: when, precisely, did Einstein learn of the experiment? Using the Archives we can try to answer this with some conviction. Crucial among the recent documents are a series of letters to Mileva, which incidentally also revealed the story of Lieserl. In a letter of September 1899 he spoke of reading a paper on electromagnetism and moving bodies by Wilhelm Wien, who did important work on black-body radiation that was part of the early history of quantum theory.[7] Although Einstein did not mention the details of the paper, we know that it contained the following important information relevant to this story: Wien lists thirteen experiments directed to the problem of measuring the earth's motion through the aether, with the last one on the list being Michelson's. If Einstein read the paper closely, it could have been a source of his first awareness of the experiment; but, unfortunately, this fact tells us nothing of Einstein's response to it or to the other experiments on Wien's list. But we can say that at least by 1907, Einstein not only knew of the experiment, but also referred to it in support of the relativity principle in an important research review essay that we shall discuss in the next Chapter.[8] Additionally, in a letter of 1908, he again referred to the experiment, in this case speaking of it as having "put us in the worst predicament," which was resolved by the relativity theory.[9] Nonetheless, these post-1905 remarks do not support a definitive role for Michelson's experiment in the genesis of the theory.

In 1969, long before the recent coming to light of the above documentation, science historian, Gerald Holton published a lengthy, meticulously researched article, in which he argued that Michelson's experiment had little influence on the genesis of the theory, although Einstein probably was aware of it (and we now know he was at least exposed to it), and further that it may have been one of the cases (today we would say, along with possibly twelve more) he had in mind when he wrote of "the

---

[4] Einstein [49] [1917], pp. 52–54, in Chap. 16, Experience and the Special Theory of Relativity.

[5] Importantly, even if it were true that Einstein was influenced by the experiment in relativity's formative stage, it is not a one-step jump to postulating the theory. If so, others who read Michelson's work would have published the theory before Einstein.

[6] Einstein, quoted in Stachel [190], pp. 45–46.

[7] Stachel [190], pp. 45–46; *Einstein Papers*, Vol. 1, Doc. 57.

[8] *Einstein Papers*, Vol. 2, Doc, 47. This paper will be discussed not only in the next chapter but later in this book, for other reasons.

[9] Letter of January 14, 1908 to Arnold Sommerfeld, in *Einstein Papers*, Vol. 5, Doc. 73, pp. 50–51 ET. Also, see Staley [193], p. 11.

unsuccessful attempts to detect a motion of the earth relative to the 'light medium'."[10] Holton's now-classic article, however, has sometimes erroneously been interpreted as asserting that Einstein was not at all aware of Michelson's experiment when he published his famous paper in 1905. On the contrary, the argument was much more subtle; Holton rather was demoting the role of Michelson's experiment from the premier place in the genesis of the theory as expounded in popular accounts of relativity to a minor supporting role, along with other more significant cases.

This is important to keep in mind because Holton's thesis was seriously challenged in 1982 with the publication of what was purported to be a newly-discovered essay by Einstein with the provocative title: "How I Created the Theory of Relativity."[11] This essay, in fact, was not written by Einstein, but was a transcript of a lecture by him delivered in German to students at Kyoto University in December 1922, while on a visit to Japan. He was asked to speak extemporaneously on the topic. The translator then rendered Einstein's talk into Japanese in his (i.e., the translator's) own words, and this was later translated into English.[12] The nature of the documentation must be kept in mind when approaching this essay, since it is not a valid primary source in the usual sense in which historians use the term.

The crucial passage in the text followed a section where Einstein mentioned the experiment he conceived of as a student to detect the motion of the Earth through the aether, which we now know he actually proposed, and where he then goes on to mention that he knew of Michelson's work at the time. He is quoted as saying this:

> While I had these ideas in mind as a student, I came to know the strange result of Michelson's experiment. Then I came to realize intuitively that, if we admit this as a fact, it must be our mistake to think of the movement of the earth against aether. That was the first route that led me to what we now call the principle of special relativity.[13]

The document, however spurious, thus places Michelson's experiment within the genesis of relativity.

Holton's response in 1988 was to question how closely the translation was faithful to "what Einstein may have said in 1922."[14] Serious questions on the legitimacy of this essay were also raised by Arthur I. Miller, another well-known Einstein scholar.[15] He noted, among other things, that Einstein was never sent a copy of the transcript for approval; as well, there is no mention in the essay of factors that we know were part of the genesis of the theory, such as Faraday's experiment or the thought experiment of riding a beam of light. Agreeing with these scholars, I was

[10] Holton [99] [1969], pp. 279–370. The title, incidentally is, "Einstein, Michelson, and the 'Crucial' Experiment."

[11] Einstein [52] [1922].

[12] Einstein [52] [1922]; Ogawa [156]; Abiko [1].

[13] Quoted in Abiko [1], p. 13. Abiko is critical of earlier translations such as that of Ogawa [156], pp. 79–80, and Ono, See Einstein [52], p. 46.

[14] Holton [99], p. 480.

[15] Miller [142].

surprised at the time to find Einstein's otherwise excellent biographer, Abraham Pais, whom I highly respect, buy into the authenticity of the "document."[16]

Until recently, this document was undeniably a puzzle, since apparently nowhere else (before or since) did Einstein place Michelson's experiment in this key role at the start of his theory. Subsequently, while writing this book, another document came to light in the most recent volume of Einstein's papers. It is another case of a transcript of a lecture by Einstein, using notes from an unknown person in an audience. This one is dated 1921 in Chicago, when Einstein was on his first visit to the United States, primarily in support of the Zionist movement, which specifically was raising funds for the creation of a Hebrew University in Jerusalem. During the year's tour he also delivered lectures on relativity in various places across the country. In May, at a school in Chicago,[17] he spoke of how he came to the theory of relativity and is quoted as saying he knew Michelson's experiment "when I was a student."[18] Dated one year before the Kyoto lecture, this document reinforces the possible validity of both documents. Thus it appears true the Einstein did say in the early 1920s that he knew of Michelson's experiment about the time of his writing his 1905 paper on relativity. In spite of this substantiation of his pre-1905 awareness of Michelson's experiment, we may still agree with Holton that "the evidence is that the influence of the famous experiment was neither direct and crucial nor completely absent, but small and indirect."[19] That would explain why he did not mention it specifically in 1905.

Finally, we should keep this in mind: the one, and only one, experiment Einstein referred to explicitly in the 1905 paper was Faraday's on the motion of a magnet and a conductor.

6.1   A Note to the Reader

At this juncture in the book we must depart from a close reading of Einstein's exact texts because of their often technical and sometimes advanced mathematical nature: this will hold true for most of the remainder of the book. Nonetheless, I will try to explain the theories as clearly and simply as possible, to hold as faithfully as possible to their original context, and not dumb-down the material, thus robbing it of its conceptual and physical complexity. When feasible, quotations will come from original sources; as well, the text, at times, will not shy away from some basis mathematical expressions.

---

[16] Pais [162], pp. 116–119 and 172–173.

[17] The school was the Francis W. Parker School, a K–12 school founded in 1901 based on John Dewey's principles of progressive education. This was much in line with Einstein's liberal views of learning, and probably why he was asked – and he accepted – the invitation. The private school is still thriving near Lincoln Park in Chicago.

[18] Van Dongen [204]. The transcript specifically reads: "… als ich im Studium war": that is, "when I was in studies …" – implying, it seems, that he was a student when he learnt of Michelson's experiment. *Einstein Papers*, Vol. 12, Appendix D, transcript on p. 519. See introductory note on p. 513.

[19] Holton [99], pp. 479–480.

Chapter 7
What Does the Theory Predict?

Einstein, having eliminated the aether – specifically dismissing it as superficial – proceeded to deduce a world based entirely on his two postulates. The first topic in this world of Einstein was the concept of time. Newton, recall, very acutely pondered mass and time, and he concluded that measured time was relative, whereas the time of passing events (such as our rate of aging) must be absolute. But Einstein deduced something quite different. Here is a simple visual example showing what he found.

Consider a train moving straight across an open space and passing an embankment. One observer, Jack, is on the embankment, and his twin sister, Jill, is in the train. The train is moving at a constant speed (**v**) and thus the train is an inertial system for Jill. Jack, being at rest on the embankment is in an inertial system too. Specifically, Jill is exactly in the middle of a boxcar that has mirrors at each end (Fig. 7.1). She controls a light source that sends two simultaneous beams of light toward the two mirrors.[1] Just when Jill passes Jack she transmits the beams, which are seen by both of them, since the boxcar has an open side toward Jack. We wish to compare the experiences of the siblings, specifically when each sees the light beams arrive at mirror **F** (at the front) and mirror **B** (at the back).

For Jill everything behaves as if she were at rest, since she is in an inertial system. So by the principle of relativity the two beams strike the mirrors at the same time; for her the two events are simultaneous. To understand what Jack sees, it is helpful to analyze the two light beams in slow-motion. From Einstein's second postulate, the beams move independently of the speed of the train, and so both beams move at speed **c** toward the front (**F**) and the back (**B**) as Jack views them; this means that the two beams move forward and backward at the same rate with respect to the landscape (and therefore with respect to Jack). While this is happening, however, the car itself is moving forward, and therefore the beam moving toward the back of the car

---

[1] This example using mirrors is from Einstein and Infeld [59] [1938], pp. 178–179. An often used example in the popular literature on relativity has the train tracks struck simultaneously by lighting at two ends; this is found in Einstein's popular book of 1917. See Einstein [49] [1917], pp. 21–27.

D.R. Topper, *How Einstein Created Relativity out of Physics and Astronomy*, Astrophysics and Space Science Library 394, DOI 10.1007/978-1-4614-4782-5_7,
© Springer Science+Business Media New York 2013

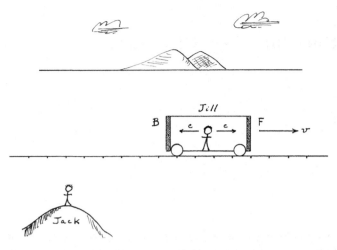

Fig. 7.1 Einstein's thought experiment on the relativity of time. The illustration involves two observers, Jack on the embankment and Jill on the train. To Jill the two light beams arrive simultaneously at the two mirrors. To Jack the beam arriving at **B** precedes the beam at **F**. So Jack and Jill measure different time intervals

will strike the back mirror before the beam moving toward the front strikes that mirror. Jack will see the one beam strike **B** before the other strikes **F**, whereas for Jill the two beams strike **B** and **F** at the same time. In short, what is simultaneous for Jill is not simultaneous for Jack. Since simultaneity is the basis of the concept of absolute time then, Einstein reasoned, time is relative for Jack and Jill.

Before we look closer at this crucial deduction, it worth noting the key role the second postulate plays. If the speed of light were not independent of the motion of the source, the beam moving toward the front would travel at the speed of light plus the speed of the train (or $\mathbf{c} + \mathbf{v}$) and the beam moving toward the back would travel at the speed of light minus the speed of the train (or $\mathbf{c} - \mathbf{v}$), so that Jack too would see the beams striking **B** and **F** as simultaneous events. This example plainly shows how the second postulate of the invariance of the speed of light leads to the relativity of time.

Einstein's actual discussion of time in the 1905 paper was expressed in experiential language. He wrote: "If, for example, I say that 'the train arrives here at 7 o'clock,' that means, more or less, 'the pointing of the small hand of my watch to 7 and the arrival of the train are simultaneous events'."[2] This set the definition of local time as the time recorded by a local clock. (Later, Chap. 8, we will see Peter Galison's analysis of Einstein's definition of time within the context of his work at the Patent office.) Using the train example we give Jack and Jill each identical clocks, which were synchronized when then they were previously together at rest. When the train passes

---

[2]Leopold Infeld once called this "the simplest sentence I have ever encountered in a scientific paper." Infeld [107], p. 27.

Jack, we saw that Jack sees event **B** before event **F**, whereas Jill sees the two events as simultaneous. Interpreting this in terms of the two clocks, one each for Jack and Jill, Einstein deduced (although he was not speaking of Jack and Jill) that Jack sees Jill's clock as running slower than his. Working-out quantitatively the time-difference in the running of the two clocks, he derived the following result: the moving clock slows down by the square root of the quantity: **1** minus the square of the train's speed over the square of the speed of light. Mathematically it is:

$$\sqrt{(1 - v^2/c^2)}.$$

Because this is a term that appears frequently in special relativity it is convenient to give it its own designation: I like to call it **Q** (for, the quantity). When applied to the concept of time, **Q** was later known as the "time dilation," a number measuring how much the time of a moving clock slows-down relative to a clock at rest. This result is, in essence, the relativity of time, which in 1905 Einstein said was a "peculiar consequence" of the theory.[3] Peculiar, indeed.

Before looking more closely at the relativity of time, it is important to bring up an erroneous presumption about the relativity of time that is found in some popular writings (and even some non-popular one). There is no time-reversal in relativity; that is, it is not true that events in one system may be seen as reversed in another system (where the 'after' comes before the 'before'). This means, importantly, that causality (the law of cause and effect) is not violated in relativity. The sequence of causal events always proceeds in succession in all systems; only the temporal 'distances' between events may differ. Consider this example using a pool table: a pool stick strikes a blue ball, the blue ball moves 6 cm and smashes into two balls (one red and one yellow), with the red ball going 12 cm into a side pocket and the yellow ball 20 cm into a corner pocket. All viewers will see the same sequence of causal events, but due to their relative motions they may differ in their measurements of the times between the events. Thus one viewer may see the red ball fall into the side pocket before the yellow ball falls into the corner pocket; whereas another viewer moving relative to the first viewer sees the yellow ball fall before the red ball. But no one sees these non-causally connected balls fall in the pockets before the blue ball hits them. To repeat: strict causality is still obeyed, and there is no time-reversal in relativity. Contrary to a popular limerick, one cannot go on a trip in a relativistic way and return on the day before.

Since the **Q** term [$\sqrt{(1 - \mathbf{v}^2/\mathbf{c}^2)}$] plays a central quantitative role in relativity, we need to look at it more closely. Consider some specific physical cases. Start with $\mathbf{v} = 0$, with Jill at rest. Here **Q** $= 1$; which means, both Jack's time and Jill's times are the same; there is no time dilation, and the clocks are simultaneous. This is the redundant case, which must hold true if the theory is to work at all. Next, for all

---

[3] Einstein (1905), quoted in Stachel [191], p. 139. The mathematical deduction is from the 1905 paper; the train example, as noted before, is a variation of a visual example used later in his popular book.

cases of **v** much smaller than **c,** the term v^2/c^2 is near zero (since **c** is such a large number) and, again, **Q** is near unity (1). A closer look at **Q** reveals that Jill needs to get quite close to the speed of light (such that v^2/c^2 takes on a tangible value compared with 1) before **Q,** in turn, takes on a value appreciably different from unity. In the jargon of the theory, we say that relativistic effects materialize only near the speed of light. Relativity, in other words, is not a factor for most everyday experiences, where Newton's physics works just fine.[4]

To get a handle on the relativistic effect, consider a specific case. It is useful to use **v** = 98% **c**; with Jill is moving at 98% the speed of light. The math is simple: $(0.98)^2$ is very nearly 0.96, and 1−0.96 = 0.04. The square-root of 0.04 is 0.2 or 2/10 = 1/5. Hence Q = 1/5. Applying this to Jack and Jill: if Jill is traveling at 98% the speed of light, then when 5 years pass on Jack's clock, he will record only 1 year on Jill's. Her clock is running slower than his by 1/5. Taking this further: the closer she gets to **c,** the slower the time-rate, with the limiting case being that time itself (that is, the passing of time) would stop (**Q** = 0) at the speed of light. This means that the speed of light is a physical limit for anything moving; and, further, since matter cannot reach the speed of light, it cannot exceed light-speed, either.

In many ways the time dilation is the strangest deduction of special relativity, since it violates our (and Newton's) intuitive sense of time. In 1905 Einstein interpreted this result as meaning that a moving clock does run slower than a clock at rest, and that it will, after moving a given distance (call it **d**), record a slower time than the rest-clock. He further considered the clock moving these same **d** intervals along a polygon with sides of the same length **d**, and concluded that the clock would continue to run slowly even when it made a complete cycle, retuning to the rest-clock. Einstein, who did not shy away from making intuitive generalizations, then asserted: "If we assume that the result proved for a polygonal line holds also for a continuously curved line" then after making a round trip, and comparing the rest-clock with the moving-clock, the latter "clock will lag…behind the clock that has not been moved."[5] He did not flinch at this "peculiar consequence," although later (to be seen in Chap. 9) it was used by others as the basis for an apparent paradox. What is more, he carried the logic further: if the theory really did mean that time itself slows down – not just the rate of clocks – then it would also mean that our biological clock (as we call it today) would slow down as we approach the speed of light. That our rate of aging was a relative quantity, not surprisingly, became a matter of considerable debate.

After deducing the time dilation, Einstein made another strange deduction from the two postulates of relativity – a length contraction. Applying what he found to our example: if Jill held a meter stick parallel to the train tracks, Jack would see it shrink compared to his identical meter stick, and the amount of contraction was the same quality **Q.** As with the time dilation, this effect was manifest only at speeds

---

[4]We will see later that this is not entirely true. But bear with me until we get to contemporary examples.

[5]Einstein (1905), quoted in Stachel [191], p. 139.

approaching light-speed. For Jill at rest, her meter stick was the same length as Jack's. But if she traveled at 98% the speed of light, Jack measured her meter stick to be 20 cm (1/5th of a meter). Jill, nonetheless observed no contraction. As with the time dilation, light-speed again was a limiting case, since at that speed the object would shrink to zero length, or disappears.

The idea of a contraction phenomenon for something moving fast was, in fact, not unique to Einstein at this time. In the late-nineteenth century a similar idea had been put forward to explain the null result of the Michelson-Morley experiment in order to save the aether. Returning to Fig. 6.1: the motion to **B** and back taking less time than to and from **A**. Working backwards from the experimental data, if the apparatus parallel to the direction of the aether wind contracts by a minute amount (assuming, say, some small pressure on the apparatus due to the aether) this could account for the light-beams returning at the same time, since the **A** trip would be slightly shorter in distance and therefore in time. Interestingly, or significantly, the amount of contraction was calculated to be the same quantity **Q** that Einstein later arrived at deductively.

How therefore is one to interpret Einstein's result, in comparison with the previous inductive contraction hypothesis? From an empirical viewpoint this conflation of two ideas of contraction is a case of the same result arrived at by two different routes. From a realistic view of a scientific theory,[6] however, they are very different: explaining the contraction as due to an aether pressure is a viable mechanical model, in the tradition of the seventeenth century "mechanical philosophy" (as it was called). On the other hand, Einstein's framework rejected the existence of the aether, and hence there was nothing in space to impart a pressure that would cause such a contraction; in his theory the object just contracted as it moved, period. Nothing was pushing or squeezing it. The same interpretation was later applied to the time dilation too, in that the slowing of clocks was not a mechanical process; the clock's internal parts were not slowing down by some external pressure (think of a pendulum clock, with the swing going slower due to air); rather the duration of time itself was proceeding at a slower pace as the object moved. Clocks measured the pace of time. All this shows that the theory of relativity was as often said, a different way of thinking – an imaginary leap similar in ways to the change in mental framework that we saw was vital to fathom the concept of inertia.

A third deduction from the two postulates challenged the concept of mass. Conceived and named by Newton as a way around the relativity of weight (Chap. 4), mass for Einstein was, on the contrary, a relative entity likewise dependent on motion. Specifically, he deduced that as a body moves, its mass increases. Yes, increases, not decreases – unlike length contracting. For this increase in mass the quantitative factor was again **Q**. Specifically the mass increases by 1/**Q**. With Jill at rest, the mass she measures was what Einstein later called the "rest mass." When she moved with some speed **v**, her mass (and, as well, her weight too) increases by 1/**Q**, as measured by Jack. If she travels at 98% light-speed her mass is five-times her rest

---

[6]Or maybe more precisely said: conceiving of a scientific theory as a model of the world.

mass (the inverse of the 1/5). So if Jill holds a 1 kg mass, while traveling at 98% **c**, then Jack measures it as 5 kg. So much for Newton's absolute entity. Furthermore, as in cases of time and length, the speed of light **c** was the upper limit, since at that speed any mass becomes infinite – another physical impossibility.

As with the time dilation and length contraction, this increase in mass did not have a mechanical cause, such as being due to a drag on the aether (since, we know, there was no aether for Einstein). The mass, as a measured parameter, increased solely due to the nature of space and time. Time slowed, length contracted, and mass increased: these were the first physical deductions of special relativity, as put forward in Einstein's first paper on the theory in 1905.

Although we are using the term special relativity, in 1905 the theory had no such name; indeed, the paper from which this theory came was, "On the Electrodynamics of Moving Bodies." The discussions of time, length, and mass are only a part of the paper (at the beginning), but obviously the topic essential to this book. Nevertheless, let me comment briefly on some of the rest of the paper. Keeping in mind the title of the paper, Maxwell's equations provided a description of the behavior of electromagnetism. Applying his two postulates to Maxwell's equations, Einstein showed that they retain their same form in any inertial system (or, the equations are of the same for both Jack and Jill). The mathematical term for this identification of form (mentioned in Chap. 1) is covariance (similar to the invariance of terms or constants, such as the speed of light).

The covariance of Maxwell's equations for all inertial systems resolved the paradox at the start of the paper, from a mathematical point of view. As he reminded the reader: "… the asymmetry in the treatment of currents produced by the relative motion of a magnet and a conductor, mentioned in the introduction, disappears."[7] Said otherwise, the covariance of Maxwell's equations under the relativity postulates is the mathematical expression of the symmetrical reading of Faraday's experiment. This then bears-out what he wrote at the start of the paper, which we quote again: "These two postulates suffice for the attainment of a simple and consistent electrodynamics of moving bodies based on Maxwell's theory for bodies at rest." "Simple and consistent" are aesthetic terms having there conceptual and mathematical correspondence with the terms invariant and covariant. The argument, in a nutshell, had come full circle.

The paper ended with a thank you (the only personal acknowledgement in the entire article) to his friend Besso, to whom he was "indebted …. for several valuable suggestions."[8] What were these suggestions? In the next chapter we will consider this question.

***

This first paper on what became the theory of relativity was shortly followed by a three-page paper in September, with the questioning title: "Does the Inertia of a Body depend upon its Energy-content?" It was a supplement to the June paper, added as a

---

[7] Einstein (1905), quoted in Stachel [191], p. 146.
[8] Einstein (1905), quoted in Stachel [191], p. 159.

sort of afterthought. Having deduced that the mass of a moving body increased with speed, Einstein realized that the greater the mass of a body the greater its energy (we know, for example, the amount of kinetic energy is proportional to its mass). He then deduced in this short paper that the increase in mass (**m**) resulted in the following increase in energy (call it **E**): the energy divided by the speed of light squared, or **E/c²**. From this he inferred "the more general conclusion" (note, again Einstein is making an intuitive generalized leap) that the "mass of the body is a measure of its energy content."[9] In other words, the answer to the question in the paper's title was, "Yes."

Although he did not write this result as an actual equation at this time, it would be, in today's notation, $E = mc^2$; that is, mass and energy are convertible, with one transforming into the other. This added another conversion to those found in the previous century. It also meant that the concept energy, or more specifically, the law of the conservation of energy, had to be modified. Writing Einstein's equation as $c^2 = E/m$ is instructive, in that it shows that although energy and mass are relative entities, their combination (later called mass-energy) is absolute, since the speed of light is invariant.

As noted before, in 1905 he used **L** for energy and **V** for the speed of light, and so the notation he used was **L/V²** for the mass. Or, said another way, the "famous" equation he did not write was, $L = mV^2$.

***

7.1 Summary

From the two postulates of relativity Einstein deduced a world with the following phenomena (for objects traveling near the speed of light):

1. Clocks (and hence time itself) slow down,
2. Lengths contract,
3. Mass increases,
   – and all change by the same quantity, $Q = \sqrt{(1 - v^2/c^2)}$
     A further deduction is the equation
4. $E = mc^2$.

Whereas for Newton, time, length, and mass were absolute quantities, they were not so according to relativity theory.

   The conservation of energy, discovered in the nineteenth century, implied that energy was an absolute quantity, conserved in all physical and chemical processes. But Einstein's equation predicted that mass and energy are convertible. What is conserved is mass-energy.

***

---

[9]Einstein (1905), quoted in Stachel [191], p. 164. For a simple derivation on an undergraduate student's level, see French [68], pp. 319–321.

Is this imaginary world that Einstein deduced in 1905 the real world? Said another way: independently of the aesthetics of the theory – its elimination of asymmetries, its invariance of light-speed, of mass-energy, the covariance of Maxwell's equations, and maybe more – was there any experimental proof of the theory in the first decade of the twentieth century? Unfortunately, most of the deductions were un-testable at the time. The conversion of mass and energy could not be tested, and there was no way of detecting the time dilation or the length contraction – all were beyond to range of experimentation (essentially, the Q number was too close to unity). Only the mass increase was provable. To understand why, we need to go back a decade or so to the discovery of the electron.

Consider an old computer monitor or an old TV set. The origin of the picture tube goes back to experiments performed in the late-nineteenth century involving electricity in vacuum tubes. The set of experiments of interest here were performed by the Englishman, J. J. Thomson, published in 1897, which constituted the discovery of the electron. In terms of the later picture tube, electron beams were shot at a screen, illuminating it, and by controlling these electrons with magnetic fields, images were produced on the screen. The image came much later, but Thomson did measure the mass of these beams, and concluded that they were composed of charged particles, not light-waves, in the tube.[10] What was remarkable at the time of the discovery was that the measured mass of this electron was much smaller than what was predicted (from chemistry, for example) for the mass of a hypothetical atom. As seen, the very existence of atoms was still much in dispute at this time: many physicists were skeptical of atoms existing at all, although some were making progress, particularly in gas theory, by conceiving of a gas as being made up of many particles colliding with each other. Most chemists, however, viewed the atom as merely a hypothetical or metaphysical entity, since they believed that chemistry should only to be studied in term of empirical parameters, such as weight and volume. As well, an atom (as the name implied, from the Greeks) was supposedly an "indivisible" entity, by definition; surprisingly, however, the particle discovered by Thomson was about 1,000 times smaller than the predicted (hypothetical) atom. It is a rather bizarre historical fact that the first "atomic" particle found was, in fact, sub-atomic. After Thomson's discovery much experimental effort went into studying electrons, and decades later the TV was invented.[11]

The experimental work relevant to Einstein's 1905 prediction of an increase in mass was that performed by Walter Kaufmann in Germany. It turned out that electrons in these proto-picture tubes were moving near the speed of light, and this could be used to test Einstein's theory. Kaufmann worked on these experiments for several years and published his results in 1906; interestingly, his was the first

---

[10]Thomson initially called the particles corpuscles, but the term electron was coined by someone else and it stuck.

[11]The specific context of this discovery is much more complex that this short exposition. For a lively account that even questions Thomson's sole credit for the discovery, see Rothman [175], Chap. 6.

publication to make reference to Einstein's 1905 paper. Kaufmann found that the mass of the electrons was indeed increasing as they approached the speed of light. Qualitatively, then, his experiments appeared to confirm Einstein's theory. Kaufmann, however, did not interpret his results this way; instead he said his experiments contradicted Einstein's theory because the number he got for the mass-increase differed from Einstein's.[12] Additionally, his reason for rejecting Einstein's deduction was not only the quantitative difference in mass, but it was primarily based on a different physical interpretation of the change in mass. Kaufmann asserted that the increase in mass that he measured was due to an aether drag; the electrons were moving so fast that they dragged some aether with them, such that his experiments were measuring both the real mass of the electrons and an extra aether drag. Believing steadfastly in the existence of an aether (as most physicists still did), Kaufmann was sure that an electron's intrinsic mass was a constant. Einstein's new way of thinking made no sense to him.[13] Therefore, the first published response to Einstein's theory was negative. What was Einstein's reaction? To answer this requires a short digression.

Einstein did not become famous in 1905. There was no wide-spread reaction to relativity theory. But a handful of physicists found it intriguing. One, who later turned against the theory and even Einstein himself, was the German physicist, Johannes Stark.[14] He was editor of a yearbook of physics and in 1907 he asked Einstein to write a review article of the theory. This request alone shows that at least he took the theory seriously at the time. Einstein worked enthusiastically on this project in his spare time, producing a long paper, in which, as he wrote to Stark, he tried to emphasize "the intuitiveness and simplicity of the mathematical developments" so as to "make the paper more stimulating." The article appeared in the 1907 issue of Stark's journal as, "On the Relativity Principle and the Conclusions Drawn from It."[15]

In their correspondence during the writing of this article, Stark alerted Einstein to a study of Kaufmann's work and Einstein responded in the review article with this statement: "Only after a more diverse[16] body of observations becomes available will it be possible to decide with confidence whether the systematic deviations [in Kaufmann's experimental results] are due to a not yet recognized source of errors or to the circumstance that the foundations of the relativity theory do not correspond with the facts."[17] The position Einstein articulated in this sentence was consistent with other statements on the role of experiment in his thinking around this time.

---

[12] What I have called **Q** in this book.

[13] Topper [198], pp. 8–9.

[14] He went on to win the 1919 Nobel Prize in physics for a range of experimental work.

[15] *Einstein Papers*, Vol. 2, Doc. 47. It was actually published in January of 1908, but historians speak of it as the 1907 relativity article, since it appeared in the 1907 issue. In essence, this research review article is the first history of the theory of relativity. As noted in the previous Chapter, at the start of the paper he explicitly mentions the Michelson-Morley experiment for empirical support.

[16] *Mannigfaltigeres*, in German, or more manifold.

[17] *Einstein Papers*, Vol. 2, Doc. 47, p. 461 and p. 283 ET.

Contrary to some assertions on this topic, Einstein did not dismiss experimentation
in physics so as to bolster the role of aesthetic factors. Instead, he believed that one
experiment alone could not negate or falsify a viable theory; rather a manifold,
diversity, or variety of them was required either to disprove or to support a theory.
I am here reminded of the phrase in the 1905 paper, from which he dismissed the
aether: "Examples of this sort, together with the unsuccessful attempts [note the
plural] to detect a motion of the earth relative to the light medium ...." In the end
Einstein was right; there were systematic errors in Kaufmann's work, but this was
not understood until around 1916 when others repeated the experiments and
confirmed Einstein's theory.[18]

Let's dwell a bit more on Einstein's view of role of experiments, for the myth of
his anti-empiricism is well-entrenched in popular writing. As a student he was much
enamored by experimental work; as a consequence, he was inclined to look favor-
ably on experiments as the ultimate arbiters to make or break a hypothesis – however much aesthetic and other factors came into play. But, as noted, the emphasis,
importantly, was on the need for a cluster of experiments, not an isolated one, to test
an idea. Listen, for example, to the following sentence from the introduction to his
1905 paper on what became the quantum theory of light; the details are not of interest here, but note the list of empirical phenomena as evidence for his hypothesis.
"Indeed, it seems to me that the observation of 'blackbody radiation,' photolumines-
cence, production of cathode rays by ultraviolet light, and other related phenomena
associated with the emission or transformation of light appear more readily understood if one assumes that the energy of light is discontinuously distributed in
space."[19] Just as at the beginning of the relativity paper (recall the role of Faraday's
experiment), so in the quantum paper, empiricism played a key role in supporting
his hypothesis.[20] Einstein, I believe, developed a considerable respect for experimentation while tinkering not only in the ETH labs, but maybe too from hands-on
experience as a child in his father's electrical workshop.

***

Returning to the 1907 review article: notice that in his letter to Stark, quoted
above, Einstein used the phrase "foundations of the relativity theory."[21] He used the
phrase "the theory of relativity" also in the introduction to the article.[22] This gives
rise to the issue of the chorology of terminology; specifically when various terms
we use today were first introduced. A brief outline, as far as I have been able to
reconstruct, is this. In a paper of 1906, Einstein speaks of the 1905 paper as "my

---

[18]Holton [98], p. 10. Apparently there was a small leak in Kaufmann's vacuum system.
[19]Einstein (1905), quoted in Stachel [191], p. 178.
[20]Hentschel [88], makes a strong case for this viewpoint. See also Van Dongen [205], Chap. 4, for
further discussion of Einstein's changing attitude toward experiments.
[21]"*Grundlagen der Relativitätstheoire*" in *Einstein Papers*, Vol. 2, Doc. 47, p. 461.
[22]"*Die Relativitätstheorie*" in *Einstein Papers*, Vol. 2, Doc. 47 p. 436, p. 254ET.

paper on the principle of relativity."[23] Max Planck in a 1906 lecture used the term relative-theory[24] but another scientist (Alfred Bucherer) in a discussion following Planck's lecture first used the term theory of relativity.[25] Einstein then used it is a letter to Paul Ehrenfest in 1907 in this phrase: "If the theory of relativity is correct …,"[26] and for years he used both "principle of relativity" and "theory of relativity." In 1911 a leading mathematician, Felix Klein, proposed the nomenclature "invariant-theory," which in some ways better expressed what the theory was about; that is, replacing old absolutes with new ones. But neither Einstein nor others adopted it.[27]

Also in 1911, Einstein spoke of his 1905 theory as "the ordinary relativity theory"[28] to distinguish it from the emerging general theory. The next year, in a letter, he used the terms "special" and "general" for the two,[29] which (to be seen) he then used in his papers of 1915–1916. So we have today's terminology: the special and the general theories of relativity.

***

Finally – recalling the 1905 paper, now in the context of the discovery of the electron – I wish to point out something that was not mentioned before: the derivation of the increase in mass was initially derived for the electron alone. After this he carried-out another act of generalization when he wrote that "these results about the mass [of an electron] are also valid for ponderable material points, because a ponderable material point can be made into an electron [which has mass] … by adding to it an *arbitrarily small* electric change."[30] Here was another intuitive leap, in this instance from the mass of an electron to the mass of all matter.[31]

These deeper looks into Einstein's original papers remind us again that the theory of relativity had its origin in the "electrodynamics of moving bodies."

---

[23] *"das Relativitätsprinzip"* quoted in Pais [162], p. 165.

[24] *Relativtheorie*, quoted in Stachel [192], p. 192. Incidentally, Planck was the referee for the relativity paper in the *Annalen* and thus was responsible for its publication.

[25] *Relativitätstheoire*, quoted in Stachel [192], p. 192.

[26] "Falls die Relativitätstheorie zutrifft…" in *Einstein Papers*, Vol. 2, Doc. 44, p. 411 and p. 237 ET.

[27] Stachel [192], p. 192. As late as 1921 Einstein received a letter from an engineer who proposed the term invariant-theory. *Einstein Papers*, Vol. 12, Doc. 250. Einstein replied: "Now to the name relativity theory. I admit that it is unfortunate, and has given occasion to philosophical misunderstanding …. The description you proposed would perhaps be better; but I believe it would cause confusion to change the generally accepted name after all this time." Quoted in Holton [100], p. 132.

[28] Quoted in Pais [162], p. 195.

[29] Pais [162], p. 210.

[30] Einstein (1905), quoted in Stachel [191], p. 157.

[31] One possible puzzle that later arose from relativity theory was this: how can light travel at the speed of light and not have infinite mass? After all, in Einstein's other paper on what became the photon (quantum) theory of light, these light-particles have mass. The answer, or more correctly the finesse around this puzzle, was to say that the rest-mass of a photon is zero, and thus its mass at **c** is not infinite. But photons are never at rest, anyway.

Chapter 8
From Railroad Time to Space-Time

In a brilliant piece of historical sleuthing Peter Galison placed Einstein's 1905 analysis of the simultaneity of time in its cultural context.[1] In particular, he showed how Einstein's work at the Patent office spilled over into his physics research.

With the rise of the railroad in the late-nineteenth century, the problem of clock coordination between cities and towns became of critical importance, which was just around the time when Einstein was a student. Not only was it necessary for the trains to run on time; but, since they ran in both directions on single tracks, the synchronization of clocks was imperative to prevent collisions.

All this had special relevance for Switzerland's clock-making industry. Bern, where Einstein worked in the Patent office, had clocks running at the same time on street corners, office buildings, and the railroad station, which he must have observed on his way to work. At the Patent office he was assigned specifically to electromagnetic patents; this meant that he analyzed numerous patent applications for electric devices related to clocks and time-related gizmos to coordinate time-signals. Unfortunately for historians today, almost all documents related to his work were destroyed long ago, this being standard procedure for the clearing of a backlog of files every eighteen years. Only a few of his documents were not destroyed because they were used in court cases where he was called as an expert witness, revealing that, in Galison's words, "Einstein soon became one of the most esteemed technical authorities in the patent office."[2]

Importantly, Galison's work casts considerable light on Einstein's paper "On the Electrodynamics of Moving Bodies" with its seemingly out-of-place experiential language in his definition of time ("the pointing of the small hand of my watch to 7 and the arrival of the train are simultaneous events") and the accompanying extensive discussion of the problem of the simultaneity of clocks – which we may now perceive less within an abstract philosophical framework and more from viewpoint of contemporary practical and technical problems. As well, Galison points out that

---

[1] Galison [74, 75].

[2] Galison [76], p. 69.

D.R. Topper, *How Einstein Created Relativity out of Physics and Astronomy*, Astrophysics and Space Science Library 394, DOI 10.1007/978-1-4614-4782-5_8,
© Springer Science+Business Media New York 2013

the style of the paper – its lack of extensive footnotes, few citations, and a focus on "simple physical processes that seem far removed from the frontiers of science" – was closer to a patent application than a typical physics publication.[3] What this reveals is that Einstein's famous paper was at once grounded in the concrete world of electricity and machines (this being his father's world, in which Albert grew-up, and out of which Hermann hoped his son would be an engineer); while, that the same time, his relativity paper soared into flights of abstract philosophical themes involving time, space, and matter. Galison nicely sums-up his journey into the historical context of Einstein's paper this way: "We find metaphysics in machines, and machines in metaphysics."[4]

Clock coordination was also central to another issue at the time – the accurate measurement of longitude and its application to setting time-zones. This problem was particularly important to the French mathematician-physicist-philosopher Henri Poincaré.[5] He was heavily involved with administrative matters in French research into not only longitude, but measurement standards, mapmaking, mine safety (he was a mining engineer early in his career), and other matters central to the French empire.[6] He was the author of an influential book on the philosophy of science, *Science and Hypothesis* (1902),[7] which, incidentally, Einstein's Olympia Academy read.[8] In it was an idea of the relativity of motion. At the time, Poincaré was famous as an author, scientist, and bureaucrat. In fact, at a scientific congress at the World's Fair in St Louis, Missouri, in 1904, Poincaré gave a speech proposing that the "principle of relativity" will solve some of the present crises in physics.[9]

Attempts to discredit Einstein's role in relativity often point to the idea of the relativity of motion as put forward by Poincaré. The scholarly claim was made by Sir Edmund Whittaker in 1953,[10] who specifically pointed to the St. Louis lecture. But a careful reading of that speech, along with other writings of Poincaré, reveals several crucial differences between his and Einstein's ideas of relativity. Poincaré's principle was based on the aether, such that the speed of light is measured as a constant number only with respect to the aether at rest; as well, he did not extend the concept of local time measurement for clocks to all systems, as Einstein did.[11] Although Einstein, as seen, only referred to his quantum theory of light as being

[3]Galison [74], p. 385; Galison [75], pp. 255–256.

[4]Galison [74], p. 389.

[5](1854–1912).

[6]Galison [75], pp. 211–212.

[7]Poincaré [168] [1902].

[8]Solovine, in his introduction to Einstein's letters, says that this book "engrossed us and held us spellbound for weeks ...." Einstein [54], p. 9.

[9]Poincaré [167] [1904].

[10]Whittaker [210] [1953]. See Chap. 2, especially pp. 30–31, on the 1904 lecture. Interestingly, Whittaker was the author of Einstein's obituary-biography for the Royal Society, where he repeated his claim. Whittaker [208], pp. 40–43.

[11]Miller [143], p. 172 in the 1981 edition; Galison [74], p. 374; and Dyson [36].

"very revolutionary,"[12] in fact, the contrast between his relativity theory and Poincaré's demonstrates how revolutionary it also was, especially when compared to Poincare's conservatism. Yet, as will be seen, Whittaker's thesis was not an isolated case of using Poincare's principle as a means of questioning the originality of Einstein's contribution.[13]

Related to this work on clock coordination was a key insight of Einstein that led to the theory of relativity and that has been dated as May 1905.[14] Apparently the catalyst was a long walk on a beautiful day with his friend at the patent office Besso, during which Einstein pored forth his perplexity over a problem that was bothering him for many years. What Besso said, or how much he even contributed to the conversation, is not known. But when they met the next day Einstein is quoted as immediately thanking Besso for helping solve the problem. The answer lay in the rejection of absolute time (along with the previous rejection of absolute space), thus completely solving "the previous extraordinary difficulty ... for the first time."[15] By the end of June Einstein wrote up a draft of his first paper on what became the special theory of relativity. This story (although we don't know the details) may hold the clue to why at the end of the paper, as noted before, he thanked only one person: "my friend and colleague M. Besso" to whom "I am indebted...for several valuable suggestions."[16]

This story may also have some relevance to the resolution of the emotional anxiety he later said he experienced in formulating the theory, as quote in the epigraph to Part II of this book: "... I must confess that at the beginning, when special relativity began to germinate in me, I was visited by all sorts of nervous conflicts."[17]

***

In 1905 Einstein was an unknown patent clerk in Bern with a Ph.D. The appreciation and acceptance of what was to become Einstein's special theory of relativity was slow and gradual. Initial interest centered on issues around the electrodynamics of the electron, since this was a key topic in physics at that time. The 1907 review article in Stark's journal should have made a contribution to its dissemination. Yet the first textbook on the theory did not appear until 1911, written by the German physicist, Max von Laue, who later won a Nobel Prize for his experimental work on shooting x-rays through crystals.[18]

---

[12] In his 1905 letter to Habicht; *Einstein Papers*, Vol. 5, Docs. 27 and 28.

[13] The claim against Einstein's originality has recently been raised by some ultra-conservative American groups. See Chap. 16.

[14] Galison [75], pp. 253–255; Fölsing [65], pp. 155–156; Abiko [1], p. 14; Editorial Note: "Einstein on the Theory of Relativity," *Einstein Papers*, Vol.2, pp. 253–274, see p. 264.

[15] Quoted in Abiko [1], p. 14, although I acknowledge the limitations of this as an exact translation of Einstein's Kyoto lecture of 1922. See previous comments on this source, above, regarding the Michelson-Morley experiment.

[16] Einstein (1905), quoted in Stachel [191], p. 159.

[17] Einstein, 1919/20, quoted in Moszkowski [147], p. 4.

[18] Von Laue (1879–1960) was a student assistant to Planck and learnt of relativity through him. (As noted in a footnote above, Planck was the referee for the relativity paper in the *Annalen*.) Von Laue once went to Bern to speak with Einstein about relativity, and was surprised to see they were the same age. They remained friends thereafter. He later wrote a further book on general relativity. For Von Laue's first book on special relativity, *Das Relativitätsprinzip*, see Staley [193], pp. 334–339.

The broader context of the significance of the theory was given a major boost in 1908, with a famous lecture delivered to the Assembly of Natural Scientists and Physicians in Cologne, Germany by Hermann Minkowski, titled "Space and Time."[19] Today, in the popular mind, one common notion about the theory of relativity is the idea expressed in the phase, "time is the fourth dimension." This is not quite true, but it is true that time did eventually appear in the theory through the fourth dimension, and this was put forward by Minkowski in his lecture.

Born in Russia, Minkowski[20] obtained his degrees in Germany, and was one of Einstein's mathematics teachers at the ETH in Zürich, whose classes Einstein often cut. Not surprisingly, Einstein made a poor impression on Minkowski, who once referred to Einstein as a "lazy dog."[21] He later was surprised at Einstein's accomplishments and found it ironic to be contributing to his relativity theory. Although Einstein was enrolled at the ETH to be a physics and mathematics teacher, he viewed mathematics primarily as a tool for physics, and had little patience for studying pure mathematics, especially if he saw it as having no potential physical application. At this time he was quite enamored with experimental physics, spending endless hours outside the classroom tinkering in the labs. This accounts, in part, for Einstein's lack of interest in Minkowski's classes during his student years.

The essence of how the variable 'time' is part of the fourth dimension of relativity theory involves fairly simple mathematical concepts. Start with Einstein's event of looking at one's watch at a particular place at a particular time. The place ($\mathbf{P}$) can be described by three dimensions ($\mathbf{x}$, $\mathbf{y}$, $\mathbf{z}$), with each variable being a measure of distance. These three axes (think of a corner of a room) place the event at a given point ($\mathbf{P}$) in space. To specify a particular event requires a fourth variable, the time ($\mathbf{t}$), since many events take place at a given point in space over time. Hence four variables ($\mathbf{x}$, $\mathbf{y}$, $\mathbf{z}$, $\mathbf{t}$) specify a specific event in space and time. But what does ($\mathbf{x}$, $\mathbf{y}$, $\mathbf{z}$, $\mathbf{t}$) represent? Or better said, how can one represent the event symbolized by ($\mathbf{x}$, $\mathbf{y}$, $\mathbf{z}$, $\mathbf{t}$) mathematically? A key problem is that $\mathbf{x}$, $\mathbf{y}$, and $\mathbf{z}$ are spatial measurements of length, and $\mathbf{t}$ is time; and different variables cannot be directly equated – in the vernacular, apples cannot equal oranges – without a constant of proportionality.

Consider the measurement of distance. In Fig. 8.1a, the distance ($\mathbf{D}$) of point $\mathbf{P}$ from $\mathbf{O}$ is specified by the expression $x^2 + y^2 + z^2$. To see where this comes from, and to understand the context of what Minkowski put forth, it is helpful to go back to basics. Think of a flightless bug on a straight wire (Fig. 8.1b). Its distance from $\mathbf{O}$ is $\mathbf{x}$; hence $\mathbf{D} = \mathbf{x}$ (or $\mathbf{D}^2 = x^2$, which is the same thing). This bug lives in a one-dimensional world. In Fig. 8.1c, another flightless bug (not drawn) lives on a flat plane; in this two-dimensional world, the distance from $\mathbf{O}$ is expressed by the equation $\mathbf{D}^2 = x^2 + y^2$, a straightforward use of Pythagoras's theorem. It follows therefore that for a flying bug in three-dimensions, $\mathbf{D}^2 = x^2 + y^2 + z^2$. Noticing the symmetry or pattern from one- through

---

[19] *Raum und Zeit*, in German; Minkowski [146].

[20] (1864–1909).

[21] Clark [26], p. 157; Isaacson [109], p. 35.

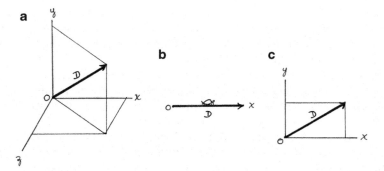

Fig. 8.1 The concept of distance in one, two, and three dimensions; i.e., (**b**), (**c**), and (**a**), respectively

three-dimensions, it follows that for four dimensions, a four-dimensional distance is expressed by the equation, $D^2 = x^2 + y^2 + z^2 + f^2$, where f is the term for the fourth dimension. I cannot draw this distance, but I can conceive of it from the mathematical expression. What now is f? Or, how do we put time (t) into f, so as to specify the event is in both space and time? The variable f has the unit of distance; so, in order to equate time with distance, time must be multiplied by a unit of distance/time (a constant of proportionality, as noted before). Now a unit of distance/time is, by definition, the concept speed or velocity; and a very likely candidate for a particular speed is c, the speed of light, which is not only a constant but also an invariant in relativity theory (from the second postulate). As a result of this reasoning, we arrive at the following possible equation for a distance in four-dimensions: $D^2 = x^2 + y^2 + z^2 + (ct)^2$.

What are we to make of this? Does it have any relevance to, or how is it connected to, the theory of relativity? Written as $D^2 = x^2 + y^2 + z^2 + (ct)^2$ it is physically meaningless; a mere mathematical expression for a four-dimensional distance, with three dimensions of space and one involving time and the speed of light. However, something in Einstein's (1905) review article provided a clue to a meaningful modification of the equation. Consider two inertial systems: (**x, y, z, t**) at rest, and (**x′, y′, z′, t′**) moving with speed **v**. Einstein deduced two equations: $x^2 + y^2 + z^2 = (ct)^2$ and $x'^2 + y'^2 + z'^2 = (ct')^2$. This meant that these equations were covariant, since they had the same form in both (and therefore all) inertial systems. Or, put slightly differently: the equations did not change their form when the observer's speed changed, or when shifting between different observers in different inertial systems. Written mathematically, the quantity $x^2 + y^2 + z^2 - (ct)^2$ was therefore covariant. This was another entity to add to the other unchanging entities in the theory, like the speed of light, mass-energy, and Maxwell's equations.[22] Invariants, remember, were crucial to the theory. Being the same in all inertial systems, invariants were part of the aesthetics that formed a foundation providing some conceptual weight (along with possible empirical evidence) for the theory.

---

[22] *Einstein Papers*, Vol. 2, Doc. 47, pp. 258–261 ET.

Minkowski then used this in a rather stunning way by writing the four-dimensional distance as $\mathbf{D}^2 = x^2 + y^2 + z^2 - (\mathbf{ct})^2$. This made the expression covariant; or put another way, $\mathbf{D}$, an invariant quantity, is a space-time distance. And this is how time entered the fourth-dimension. As the equation shows, to say that time is the fourth-dimension is as erroneous as saying that the speed of light is the fourth-dimension. They both play a role in the fourth-dimensional term. Also, rather strangely, the fourth-dimension, as a mathematical expression, is subtracted from (not added to) the spatial dimensions. But, it had to be so, for otherwise the equation was not covariant.

The invariance of the space-time distance, Minkowski believed, revealed a fundamental unity between space and time, as articulated in what became an often-quoted dramatic opening statement of his lecture. This was a "radical" vision of space and time, as he said: "Henceforth, space by itself, and time by itself, are doomed to fade away into mere shadows, and only a kind of union of the two will preserve independence."[23] Thus was born the notion, later popularized in science fiction, of space-time.

In a draft of the lecture, he originally wrote that "a fusion of the two [concepts, space and time] will show a free existence. I will provisionally call this the absolute (?) world...." The draft page contains numerous words struck-out and added, revealing Minkowski's struggle with clarifying his concepts.[24] His viewpoint, as put forward in the lecture, was based initially on an electromagnetic world-view (similar to Ostwald's energy world-view, as discussed briefly at the end of Chap. 4); Minkowski believed that Einstein's theory ("On the Electrodynamics of Moving Bodies") was an expression of this world-view. Minkowski was not alone in this viewpoint. For those who accepted the field as an independent reality, it was a short conceptual step to asserting that electromagnetism was the ultimate reality, not matter. Hence an electromagnetic world-view – distinct from the mechanical world-view or energeticism – was a conceptual belief among a number of physicists in the early years of the last century.[25]

Minkowski, however, believed he was extending this world-view further by adding a fourth-dimension. Just as moving from one- to two- to three-dimensions brought us closer to reality, so adding the fourth-dimension brought us to actual (ultimate) reality. So, whereas Newton thought space was absolute and time was absolute, and whereas Einstein showed each to be relative, now Minkowski was fusing the two into space-time, and this union was "the absolute," a world existing independently of the observer. In the draft (quoted above) notice that he used a question-mark after the word "absolute"; in the final lecture, however, he unabashedly wrote of "the postulate of the absolute world."[26]

---

[23] *Selbständigkeit*, in German; in Minkowski [146], p. 75.

[24] Galison [73], pp. 97–99.

[25] Walter Kaufmann, whose experiments on moving electrons were discussed above, held such a view. As he rhetorically put it: "Instead of all the fruitless attempts to explain electric phenomena mechanically, can we not try conversely to reduce mechanics to electrical processes?" Quoted in Holton [98], p. 4.

[26] Minkowski [146], p. 83.

The rather exotic nature of all this – what Minkowski correctly declared as "radical" – was surely a factor in the dissemination of the lecture, and imparted a further boost to bringing Einstein's theory to a wider scientific audience. Indeed, it probably was a major source of the spread of relativity over the next few years.[27] Sadly, Minkowski never witnessed the fruits of his thoughts; he died in 1909 in his mid-40s of a burst appendix, the same year his now-famous lecture was published.

Einstein, initially at least, was much less enamored with what Minkowski had wrought. Einstein's view of mathematics as a tool for physics seemed to be turned around. Unlike Minkowski, Einstein neither held the view that electromagnetism was the fundamental reality nor that mathematics was reality. A mathematical apparatus was an abstraction from an idealized version of reality. For him, therefore, more abstruse mathematics got further in the way of clearly revealing physical principles underlying reality. Minkowski was gumming-up, not clarifying the theory. As a matter of fact, in response to Minkowski's formulation Einstein is quoted as saying, perhaps slightly tongue in cheek: "Since the mathematicians have invaded the theory of relativity, I do not understand it myself any more."[28] He would, nevertheless, in time come to change his mind – indeed, radically change his mind – about this.[29]

***

Since Einstein's publications of 1905 made him well-known among only a small group of physicists, he kept his job at the Patent office, for no University appointment was forthcoming. In 1906 he got a promotion and a raise at the office, and also was awarded his Ph.D. by the University of Zürich. In a few years, as the significance and uniqueness of some facets of his papers were increasingly appreciated (such as his contribution to the reality of atoms and molecules), his fame grew among theoretical physicists, resulting in his first post as a salaried professor in 1909 at the University of Zürich. (A year before he lectured at the University of Bern as a *Privatdozent*, a peculiar German invention, where the teacher is paid from student fees alone. He had only a few students, one of whom was his friend Besso.[30]) He then left the Patent office, assuming the professorial role of teaching and pursuing research. Two years later (1911) he moved to the German University of Prague, a post lasting only a year (Mileva was very unhappy in the city), with Einstein accepting a position at, of all places, the ETH in Zürich, where he obtained his first university degree despite his benign hostility toward his teachers. They left Prague with no regrets; in a letter to friends he wrote: "… to live [in Prague] has serous drawbacks. No potable water. Much misery hand in hand with snobbishness and haughtiness. Class prejudices. Little genuine education. Everything byzantine and

---

[27] Walter [206], pp. 67–78. Walter's thesis needs to be qualified by Staley's work (mentioned below) showing how relativity was seen as a part of the theory of the electron from about 1905, and hence was more widely know because of it. Staley [193], Chaps. 6, 7, and 8.

[28] Quoted in Sommerfeld [189], p. 102. This is obviously a later recollection of Sommerfeld.

[29] For more on the relationship between Einstein and Minkowski, see Pyenson [169], especially pp. 80–81, 94–96, and 145–154.

[30] Hoffmann [97], p. 87.

priest-ridden."[31] He went on to object to the fact that his older son was required to attend Catholic church services. He especially complained about the paper work at the University: "The paper-pushing[32] in the office is interminable – all this, as it seems, in order to provide a semblance of justification for existence to the gang of scribblers in government offices."[33] On the other hand, while in Prague he pursued his research, especially (to be seen) important work on his germinating theory of gravity.[34]

One breath of fresh air while in Prague was his meeting Paul Ehrenfest for the first time. They knew of each others' work (recall Kuhn's thesis in Chap. 3 that they were the first to quantize the quantum theory), but they had only briefly corresponded before this actual meeting. Ehrenfest was searching for a posting, and in February, 1912 he made the trip from Vienna to Prague specifically to meet Einstein, who invited Ehrenfest to his home. They found much common ground – being similar in age with a similar secular-Jewish background – but mostly it was physics that kept them talking for several days, along with playing piano and violin together.[35] It was the beginning of a close friendship. When Einstein left Prague, he recommended Ehrenfest for his vacant job, but Ehrenfest eventually obtained a more prestigious position as professor at Leiden University in Holland. In Einstein's eulogy for Ehrenfest, after his tragic death in 1933, Einstein recalled that first meeting, writing: "Within a few hours we were true friends – as though our dreams and aspirations were meant for each other. We remained joined in close friendship until he departed his life."[36]

Einstein's whining about his stay in Prague, however, does not corroborate with the possible evidence, at least according to scholar Lewis Feuer,[37] that during the Prague years Einstein was a member of a close circle of intellectual Jewish friends (including Franz Kafka[38]) such that he renewed his connection with Judaism. The group introduced him to social and theological matters from Zionism to mysticism, and especially revised his interest in the seventeenth century Jewish philosopher, Baruch Spinoza.[39] Einstein was familiar with Spinoza, for Solovine reports that in the Olympia Academy days they read Spinoza's *Ethics*.[40]

---

[31] *Einstein Papers*, Vol. 5, Doc. 374, p. 275 ET.

[32] *Einstein Papers*, Vol. 5, Doc. 374. This is the English translation used in the Einstein papers for *tintenscheisserei*, which literally is "ink-shiting."

[33] *Einstein Papers*, Vol. 5, Doc. 374, p. 275 ET.

[34] Stachel [192], p. 237.

[35] Klein [116], pp. 175–179.

[36] Einstein [46] [1933], p. 215.

[37] Feuer [62], pp. xii–xix.

[38] Kafka was not a famous writer until after his death, since little was published during his life. I am not aware of any documented interaction between Kafka and Einstein within or beyond the Prague group.

[39] See also Frank [67], pp. 83–85.

[40] Solovine lists this book and other authors, such as Mach and Poincaré, along with Dickens and Cervantes' *Don Quixote*, in his introduction to the letters from Einstein. In Einstein [54], pp. 8–9.

Feuer, nonetheless makes it clear that Einstein's experience within this friendship circle did not result in a rekindling of a Jewish religious affinity (which he had abandoned around the age of twelve) or an attraction to Zionism (which came later in the early 1920s, with his exposure to virulent anti-Semitism in Berlin). Nonetheless some of his later references to an identification with Spinoza's ideas and his consequent musings about what sort of universe God may or may not have created could be traced back to his rekindling of Spinoza's philosophy while living in Prague – Spinoza, indeed, being the one philosopher he increasingly referred to positively for the rest of his life. Einstein biographer Jürgen Neffe, on the other hand, dismisses entirely the Prague circle, saying it had no influence on him, and uses as evidence a letter Einstein wrote to Hedwig Born (the physicist Max Born's wife) in 1916, where he seems to rebuff the group as "a small troop of unrealistic people, harking back to the Middle Ages."[41] Philip Frank, who assumed Einstein's teaching position in Prague after he left, paints a more positive picture of the groups' influence on Einstein, especially around matters of art, literature, and philosophy, although he does say that Einstein was not yet interested in Zionism at the time.[42]

Before leaving the topic of Einstein's time in Prague, there is an incident Frank recalls that may be worth retelling. Through the window in Einstein's office at the university he often wondered about the people in park below where only women walked in the morning and men in the afternoon. After inquiring about this peculiar arrangement he was informed that the park was really a garden of an insane asylum (as it was called in those politically incorrect days). Upon pointing this out to Frank, Einstein said: "Those are the madmen who do not occupy themselves with the quantum theory"[43] – a subject to which he continued making major contributions.

Einstein's move back to Zürich was extremely auspicious for his scientific work – even though it was short-lived. For his old friend Grossmann was Professor of Mathematics, and, as seen in the next chapter, he was invaluable in helping Einstein with the mathematics required for his gravity theory.

Einstein's continued production of highly original papers thrust him toward the top in the hierarchy of theoretical physicists, so that in the spring of 1913 a delegate from Berlin called on him, offering a prestigious research position that required no formal teaching duties. He could devote most of his time to research, while being in the company of some the best minds in physics. But it was a difficult decision, for

---

[41] Neffe [149], p. 311. In some ways this issue comes down to how one translates the phrase from the letter of September 8, 1916: "eine mittelalterlich anmutende Kleine Schar weltferner Menschen." *Einstein Papers*, Vol. 8A, Doc. 257. Neffe's translation is that found in the ET for the *Papers*. Interestingly, in an essay on "The Jewish Question" in the *Einstein Papers*, Vol. 7, pp. 221–236, written by the editors of that volume, they translate the same phrase of 1916 less dismissively as "a small band of impractical people that strike one as medieval." Another similar translation is the following from the Born letters (Einstein [56], p. 4): "a medieval-like band of unworldly people." The latter translations are more accommodating to Feuer's thesis. Surely, words like "unworldly" and "impractical" could apply to Einstein himself at times.

[42] Frank [67], pp. 77–85, especially pp. 83–85.

[43] Frank [67], p. 98.

it meant retuning to Germany, a country to which he had renounced his citizenship when leaving in his mid-teen years plus he was now a Swiss citizen. Yet, after much deliberation, he accepted the Berlin position.[44]

After he left his ETH job in the spring of 1914, he never formally taught again. His entire teaching career spanned a mere five years (or six, if the *Privatdozent* year is included).[45] On his first teaching job at Zürich, Mileva wrote: "My husband is very happy about his new post; he much prefers lecturing to the office work in Bern. His audiences are larger than the usual here and, I discovered in a round about way, people like him a lot."[46] From anecdotal evidence, it appears Einstein was a very good teacher – able to explain complex topics clearly and simply, often presenting the same scientific problem from mathematical, experimental, and philosophical viewpoints. He had a good rapport with most students, using his sense of humor and his sense of wonder to entertain and inspire them.[47] He was not, however, able to fit into the routine of an entire course, which required continued organizational skills. He did not like teaching topics he did not find interesting and his courses overall were often uneven. He was especially annoyed with administrative activities associated with teaching, such as preparation and testing, that cut into his thinking and writing time, along other paper work, as noted before. This may explain why he is reported to have said more than once that he did not like teaching, which contradicts what seemed to be true in the classroom.[48]

---

[44] Technically there were three positions. A professorship, without teaching duties, at the University of Berlin; a research position and membership in the Prussian Academy of Science; and a Directorship of the Kaiser Wilhelm Institute of Physics, the latter was not established until 1917.

[45] There was also the teaching and tutoring episode before the Patent job, where I quoted above (Chap. 3) from a 1901 letter saying he found teaching "exceptionally" pleasing.

[46] Marić, [136] [Winter 1909/1910], p. 101.

[47] Neffe [149], p. 159; Clark [26], pp. 170–171.

[48] Frank [67], pp. 89–91 and 116–119.

Chapter 9
1911: The Paradox About Time

Einstein's deduction of the time dilation was possibility the most difficult idea for both scientists and the lay public to fathom. It led to what was perceived by some as a paradox – usually called the clock paradox or the twin paradox. Before this paradox was invented, Einstein presented his interpretation of the time dilation in a lecture in Zürich in January 1911 (note the title): "The Theory of Relativity."[1] He spoke of the time dilation as both "peculiar" (a term he used before), and as "funny,"[2] for if a clock moved near the speed of light and returned to its place of origin, the hands of the clock hardly moved compared to an identical clock at rest at the place of origin. More importantly, he made it clear that his interpretation inferred not only that the clock slowed down proportional to its speed but that time itself advanced more slowly, so that "a living organism" aged at a slower rate than one at rest.[3] Here he explicitly affirmed what was implied in 1905 – that time itself, not just the ticking of clocks, changed. Einstein's published lecture, however, was not widely read.

The first person after Einstein to present the time dilation as entailing the actual slowing of time was the French physicist, Paul Langevin.[4] Being was one of the first scientists to embrace relativity theory, he gave seminars on it at the *Collège de France* as early as 1906, only a year after the theory was published. Langevin met Einstein at a scientific conference in Brussels in the fall of 1911 and they quickly became friends.

In April 1911, in a lecture to a philosophy conference in Bologna, Italy, Langevin presented the time dilation in terms of someone traveling in space. He considered the case of a space traveler making a round trip journey to a star at the speed of light, and taking one year per trip, to and fro; Langevin concluded that since the traveler aged two years, when returning to Earth, 200 years had passed – so the traveler, in

---

[1] *Einstein Papers*, Vol. 3, Doc. 17, pp. 340–350 ET.
[2] *"Am drolligsten wird die Sache…,"* *Einstein Papers*, Vol. 3, Doc. 17, p. 436.
[3] *Einstein Papers*, Vol. 3, Doc. 17, pp. 348–349 ET. See also Galison [75], pp. 266–267.
[4] (1872–1946).

D.R. Topper, *How Einstein Created Relativity out of Physics and Astronomy*, Astrophysics   77
and Space Science Library 394, DOI 10.1007/978-1-4614-4782-5_9,
© Springer Science+Business Media New York 2013

essence, was being transported into the future.[5] The published lecture – with its veneer of science fiction – was widely read, and the topic became known as Langevin's traveler. It, along with Minkowski's lecture of 1908, may have given further stimulus to popularizing relativity beyond the insular community of physicists, and especially to the community of philosophers.[6]

The paradoxical element in this example arose when objections were raised about the asymmetrical aging between the traveler and someone remaining on Earth. If the principle of relativity implied the equality of all inertial systems, then the traveler could just as well be considered at rest and assert that (say) its twin remaining on Earth was moving away and back, and that the Earth clock should likewise have slowed down according to the traveler. This meant that when the twins meet again there would be no real difference in aging, for otherwise the relativity principle would be violated. The essence of Langevin's response to the paradox, which was not unique to him (for example, von Laue, who wrote the first relativity textbook, also got into the fray), was simply this: the moving traveler experiences episodes of acceleration during the trip in leaving and returning to Earth, whereas the Earthling twin, at rest, experiences the continued acceleration of gravity. Their experiences are not identical, and since it is the traveler who is actually moving, than that traveler experiences the time dilation.[7] This example, however, brings acceleration and gravity into the case, and so moves the discussion from special to general relativity (the latter, the focus of the next Chapter); as such, we will leave the topic for now.[8] We shall see that this paradox arose again in the early 1920s, as Einstein became known beyond the closeted world of scientists and philosophers.

---

[5]But Einstein's theory prohibits reaching light-speed; as well, Langevin's example has the time dilation to be the ratio of 2/200 years, or Q = 1/100. Working backwards from this, I calculate the speed of his traveler as 99.994999 % of c, to six decimal places.

[6]Miller [143], p. 244.

[7]Miller [143], p, 248.

[8]As Miller has pointed out: "The literature on the clock paradox is voluminous, and increases daily" (Miller [143], 257); as such, we will not pursue this topic, beyond its relevance in Einstein's life, as in the next chapter.

Chapter 10
Is the Theory True Today?

As seen, the essential predictions of special relativity are:

1. Clocks (and hence time itself) slows down,
2. Lengths contract,
3. Mass increases, – all these changes are by the same amount, $\mathbf{Q} = \sqrt{(1 - \mathbf{v}^2/\mathbf{c}^2)}$.
   A further deduction is the equation
4. $\mathbf{E} = \mathbf{mc}^2$.

Part II of this book concludes with some confirmations of this theory up to the present.

As mentioned above, the mass increase of electrons moving near the speed of light was confirmed in 1916. To repeat: there were systematic errors in Kaufmann's 1906 work, but this was not understood until around 1916 when others replicated the experiments and confirmed Einstein's theory.

The time dilation was first confirmed in early 1940s with subatomic particles called muons, which have a very small "life-spans" before they decay into other particles. At rest, they last about 2 µs (that is, 0.000002 seconds). Beyond the laboratory, muons are produced when cosmic rays collide with the atmosphere; traveling near the speed of light, muons' life-spans were found to increase up to 20 µs, a tenfold increase. This was as the special theory predicted for the time dilation.

Experiments using atomic clocks have further confirmed the time dilation. More of this will be in the next chapter, since there is a second time dilation involving gravity; this topic also involves a delightful story about the first GPS system.

In the 1930s nuclear processes confirmed the transformation of mass into energy. Such processes were discovered for the Sun, thus explaining the ancient question of why the Sun does not burn-out; in essence it is a nuclear reactor, not a massive bonfire. Later the creation of the nuclear bomb and nuclear reactors further confirmed Einstein's equation $\mathbf{E} = \mathbf{mc}^2$.

Variations of Michelson's experiment continue to be performed using lasers, and the results remain null to extreme accuracy (4,000 times more accurate than the

D.R. Topper, *How Einstein Created Relativity out of Physics and Astronomy*, Astrophysics   79
and Space Science Library 394, DOI 10.1007/978-1-4614-4782-5_10,
© Springer Science+Business Media New York 2013

original experiment).[1] In short, no aether has been found, and, along with this, no absolute reframe of reference.

Finally, as for length contraction: there has not been an experiment yet devised to test this possible phenomenon.

***

10.1   Three Comments for the Advanced Reader

Comment #1: There was much discussion late in the last century about the reality of the increase in mass.[2] We now know that Einstein himself had doubts about using it. In a letter of 1948, from the Einstein Archives, he writes: "It is not good to introduce the concept of the [relative] ... mass of a body for which no clear definition can be given. It is better to introduce no other mass than 'the rest mass'...."[3] The essence of the problem is this. If an object is at rest, with rest mass $\mathbf{m_0}$ and Energy $\mathbf{E_0}$, we may write the equation $\mathbf{E_0 = m_0 c^2}$. For the object moving, however, the total Energy $\mathbf{E}$ is: $\mathbf{E = m_0 c^2 / \sqrt{(1 - v^2/c^2)}}$. If we then definite a relative mass $\mathbf{m}$ as $\mathbf{m_0 / \sqrt{(1 - v^2/c^2)}}$, we preserve the famous equation, $\mathbf{E = mc^2}$. The problem (the details of which are in the references in the previous footnote) is that for high-speed objects and under conditions involving gravity this definition does not always hold true. The only unique definition of mass is the rest mass, and that is why Einstein and others wanted to eliminate the concept of relative mass, even though the true energy equation became the more cumbersome: $\mathbf{E = m_0 c^2 / \sqrt{(1 - v^2/c^2)}}$. As early as the 1907 review article, Einstein introduced the rest mass as the true mass.[4] Even so Lev Okun has shown that relativistic mass may still be found throughout Einstein's writings on relativity.[5]

Comment #2: Some readers may have noticed the absence of the Lorentz[6] equations, or the Lorentz transformation. This is because I am trying to keep the mathematics to a minimum, and, as well, the topic is not essential for the level of this book. Speaking of Lorentz, however, brief mentioned was made of his contraction

---

[1] Will [211], p. 771.

[2] Adler [2]; McFarland [139]; Okun [157, 158].

[3] The letter, dated 19 June 1948, was written to Lincoln Barnett who was writing a popular book on relativity [7]. The handwritten excerpt of the letter (in German) is reproduced in Okun [157], p. 32. Einstein wrote the Forward to Barnett's book, which is dated September 10, 1948. However, on page 68 Barnett presents the increase in mass equation and the concept of relative mass (pp. 68–70), thus completely ignoring Einstein's guidance.

[4] *Einstein Papers*, Vol. 2, Doc. 47, p. 297 ET.

[5] Okun [158] presents the most thorough and exhaustive discussion of this matter that I know of. A deeper question on the validity of Einstein's various derivations of his famous equation over his life has been pursued by Ohanian [155], who challenges the rigor of all of them.

[6] Hendrik A. Lorentz (1853–1928), Dutch physicist.

Fig. 10.1 Diagram for a simple mathematical derivation of the time dilation, using a light-clock

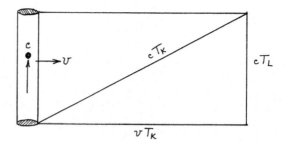

hypothesis to explain away Michelson experiment, without actually giving his name. Only passing reference was made to the fact that Einstein's relativity theory was originally seen as part of the emerging theory of the electron, which would require an understanding of Lorentz's electron theory for a deeper analysis.[7]

Comment #3: Here is a neat and simple calculation of the **Q** factor in the time dilation for those not averse to a little geometry and algebra. It uses an imaginary light-clock made by placing a photon in a cylinder with mirrors at both ends (Fig. 10.1). As the photon bounces back and forth between the mirrors it measures time (like a swigging pendulum). Put Jill at rest on the clock, and have it moving with speed **v** with respect to Jack. Call Jill's time T_L and Jack's time T_K. If the photon moves from the bottom to the top mirror, it travels a distance **c** T_L according to Jill. According to Jack the cylinder (clock) moves a horizontal distance **v** T_K and the photon moves along the diagonal a distance **c** T_K. From Fig. 10.1, and using Pythagoras's theorem, we write this equation: $(v\,T_K)^2 + (c\,T_L)^2 = (c\,T_K)^2$. Solving for T_L as a function of T_K we get this relation: $T_K = T_L/\sqrt{(1 - v^2/c^2)}$. And there it is, the **Q** factor corresponding to the time dilation, or the fact that Jack measures Jill's clock as running slower by **Q** amount.

---

[7] For a most recent analysis of this, which shows in detail how much electron theory was entwined with the early history of relativity, see Staley [193], Chaps. 6, 7, and 8.

Part III
General Relativity

...the years of anxious searching in the dark, with their intense longing, their alternations of confidence and exhaustion and the final emergence into the light – only those who have experienced it can understand it.

(Einstein, lecture in Glasgow, Scotland, 1933)[1]

Galileo used the experience of riding in a ship as an example of an inertial system. Einstein spoke of moving in a train, which was the new mode of transportation of the nineteenth century. Today we refer to riding in an airplane, which we know, or traveling in a spaceship, about which we fanaticize. (The experience of astronauts will be seen at the end of Part III.) Inertial motion is the conceptual base of special relativity. General relativity, on the other hand, is about non-inertial motion: moving from rest, slowing to a stop, or any change of speed – what we call acceleration (which includes deceleration). In Galilean terminology, this is a change of state. For this we all have the personal experience of changing one's state in an accelerating car, ship, train, plane, or any moving object.

The first goal of general relativity was to explain gravity – that mysterious force all around us and throughout the universe that Newton explained as an action-at-a-distance across empty space, a power also referred to as an occult force. The genesis of Einstein's general theory resided in a thought experiment that he later called the "happiest thought" of his life. It is found in the same 1920 document quoted in Chap. 5 on Einstein's reminisces on the origins of special relativity, where he mentioned the "unbearable" contradictions of Faraday's experiment with a magnet and a wire.[2]

---

[1] From a lecture Einstein delivered at the University of Glasgow, June 20, 1933, quoted in Stachel, 2002, p, 232, and Einstein, "Notes on the Origin of the General Theory of Relativity," reprinted in Einstein, 1954 [1933], pp. 289–290.

[2] "Fundamental Ideas and Methods of the Theory of Relativity, Presented in Their Development," *Einstein Papers*, Vol. 7, Doc. 31, p.135 ET. As noted in a previous footnote in this book: this unpublished paper (a long 35 page manuscript) from the Pierpont Morgan Library, New York, has been dated as January 1920 based on a series of letters with the editor of the journal *Nature* that almost certainly refer to this document. *Einstein papers*, Vol. 7, Doc. 31, note 3, p. 279 ET. It is doubtless the draft of an article never published. Stachel 2002, pp. 262–263.

I will scrutinize and dissect a long and important passage on the genesis of general relativity from that key document, dividing it into two parts, explaining their meaning, and placing them in their historical context. But in order to grasp fully Einstein's idea, we must first clarify the way bodies fall in a gravitational field. For this we turn again to – who else? – Galileo. (As always, the fidgety reader may skip to Chap. 12.)

Chapter 11
Galileo Discovers How Bodies Fall

While teaching at Pisa and then later at Padua, Galileo not only devised the principle of the relativity of motion, but also showed that bodies fall independently of their weight based on a thought experiment and probably an additional demonstration. The thought experiment is elegant and simple. Consider three identical weights at the same height and dropped at the same time (Fig. 11.1a). The crucial insight for this and virtually all thought experiments hereafter and right through Einstein is this: the laws of nature are to be found only in a world devoid of the resistance of a medium. Throughout the seventeenth century there was a continuing debate whether a vacuum could exist, or if the aether existed, and so forth. For Galileo, the aether, if a reality, imparted little or no resistance to motion since the planets have the same periods of motion since ancient times. His goal was to find the same regularity for the terrestrial world, and this could only be achieved if there were little or no resisting power. Assuming, therefore, no resistance by a medium in Fig. 11.1a, all three weights strike the ground at the same time. Now modify the experiment such that two of the weights are attached together with weightless glue (a product readily available in an imaginary world). Since there are no physical changes to these objects by merely combining two, they again all fall at the same rate and simultaneously strike the ground (Fig. 11.1b). This second situation, however, is different in that it can be interpreted as only two falling weights, with one twice the weight of the other. Finding no flaws in the logic of this thought experiment Galileo generalized this by inferring that all bodies fall at the same rate independently of their weight.

Like the concept of inertia, this fact about falling bodies is taken for granted today, being taught to school children along with the round and moving Earth. Yet, it too – just as the round and moving Earth – is not obvious and defies common sense. When Galileo put forth the idea, "common sense" was essentially what Aristotle had written, which was also what Galileo was paid to teach to his students. Here is an example of common sense thinking, possible what Aristotle reasoned. Consider holding two weights, a very heavy one in your right hand, and very light one in your left. Since the heavy weight is pushing harder on your right hand than

D.R. Topper, *How Einstein Created Relativity out of Physics and Astronomy*, Astrophysics and Space Science Library 394, DOI 10.1007/978-1-4614-4782-5_11, © Springer Science+Business Media New York 2013

Fig. 11.1 Galileo's thought
experiment for falling bodies.
The illustration shows that
bodies fall independently of
their weight

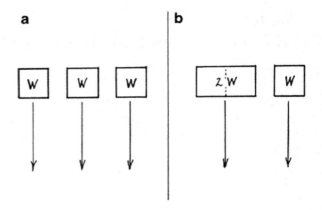

the light one on your left, it stands to reason that if you let them fall at the same time, the heavy one would fall faster, and reach the ground first. Aristotle went on to surmise that since the pushing down of the object on your hand is proportional to its numerical weight, then the speed of fall of each object should be proportional to its weight. In the case of one object being twice the weight of the other, when the heavy object strikes the ground, the light one should have fallen only half-way. So we are left with two contrasting images: two weights striking the floor at the same time, as Galileo deduced in his mind, or the lighter weight falling only half the distance as dictated by Aristotle and common sense. Which is correct?

Galileo was sure of his thought experiment but also aware of its defiance of common sense. What he did next is a matter of continuing historical debate: it involves the well-known either myth or fact around his performing an experiment with falling bodies, usually located at the leaning tower of Pisa where he was teaching at the time. We know today that in a world with a medium such as air, objects do not fall at the same rate and indeed heavier ones do strike the ground before lighter ones, which seem to prove that Galileo was wrong, while confirming Aristotle. For this reason, and since we have no direct documentation of Galileo dropping things from any tower, the Pisa story has been relegated to the realm of myth.[1] There is no consensus, however, among historians of science on this issue and I, for one, accept the Pisa story, with some modification. I do not view Galileo's probable dropping of two weights from the tower as a controlled experiment, making quantitative measurements; instead I see it as demonstration piece, showing how wrong Aristotle must be.

Let me explain this by starting with an example from today: a controlled experiment using two weights, one of 12-lbs and the other a quarter-pound. As such, the one is about 50 times heavier than the other. In dropping these weights simultaneously from 100-ft, when the heavier weight touches the ground the lighter one falls

---

[1] A recent advocate is Segre [181], who probably would not agree with my argument in the rest of this Chapter.

about 85-ft. Thus, the greater the weight the faster the fall, as Aristotle said. Does this prove Aristotle to be right? No, since a quantitative look at this experiment shows that Aristotle's rule was far off the mark, since the heavier weight should have fallen 50 times faster than the lighter one, and so the lighter weight should only have fallen about 2-ft when the heavier one struck the ground. Galileo – even using a less than controlled experiment according to today's standards – could surely have argued that the significant quantitative discrepancy between the predicted and actual differences between how the two weights fall, points toward an error of Aristotle's rule – despite the fact that the weights did not hit the ground at exactly the same time, as Galileo's thought experiment deduced.

An even closer look, I hope, seals my case. In the seventeenth-century the difference in two weights falling 100-ft and 85-ft, which we measure today using a camera, was not available. Instead the difference between them was their time of fall (not distance), and we can estimate this time difference between those two weights using the law of falling bodies. Doing so, we deduce about one-fifth of a second, which is nearly simultaneous. Further, transforming this to the possibility of Galileo doing the same experiment at the leaning tower of Pisa, which is about 185-ft high, means that the gap in time of two weights of significant difference falling from the top was only around two-fifth of a second. Clunk, clunk, … about the time it took take you to read quickly these two words out load is the difference you (as Galileo's assistant at the bottom of the tower) heard between the two weights hitting the ground. It was not a great leap for Galileo to say (I am making this up): "In the limiting case with no air resistance, the time of fall would be nearly the same for the two weights. Hence, I'm right and Aristotle and his followers are wrong." If I were Galileo's student, I would agree. Do you?

Having at least convinced himself that bodies fall independently of their weight, Galileo had to square this fact with the obvious contradiction of the experience of the heavier weight pushing down on his right hand harder than the lighter one on the left. Why does not the heavier weight fall much faster, since it is pushing down much harder? Is this not a paradox?

Inertia, Galileo's other key discovery, was the key to the answer. To see how inertia applies, we need to put forth another insight about how bodies fall without a resisting medium. Aristotle dealt with motion within a medium, since that was the only physical case he knew of, and so his falling body was assumed to fall at a constant speed. That is why he thought the speed of fall was proportional to the weight – these two being the necessary variables.[2] Galileo's bodies, however, fell without a resisting medium and because of this they were constantly speeding-up, that is, accelerating. Starting at rest and initially accelerating, nothing prevented this process from being a continuous acceleration throughout the entire time of fall. But an

---

[2] More specifically, he deduced that, since the increased resistance of a medium resulted in a slower speed, then there was an inverse proportion between the falling speed of an object and the medium's resistance. Mathematically (or quantitatively) it followed that if the resistance went to zero (that is, a vacuum) the speed would be infinite, which is impossible. This was why Aristotle always thought of motion within a medium, for a vacuum was impossible.

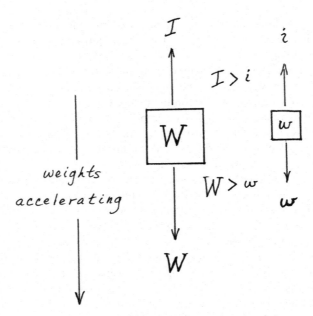

Fig. 11.2 Abstract illustration of falling bodies of different weights, and consequently different inertias. Inertia provides a resistance to a change of speed for these accelerating bodies. If the weight of each is proportional to its inertia, then they fall at the same rate

accelerating object involved a constant inertial resistance to the motion (that is, the change of speed). In Galilean terms, there was therefore a constant change of state for a falling body. Figure 11.2 is an abstract illustration of what was happening. The greater the weight (**W** > **w**) of the object, the greater was its inertial resistance (**I** > **i**) to the change of state. So, assuming that this resistance to motion was proportional to weight, then the heavier object's greater resistance to fall was proportional to that of the lighter object; they therefore balanced out, and so the two weights fell (accelerated) at the same rate – and both hit the ground at the same time. This deeper look at falling bodies, with the aid of inertia, subverted common sense and resolved the (apparent, as it turned out) paradox. So inertia became bound up with how bodies fall.

Chapter 12
1907: Einstein's Second Famous Thought Experiment

The stage is set to study Einstein's 1920 document, which we peruse posthaste. He began by mentioning the summary paper on relativity he wrote for Stark's journal:

> When I was busy (in 1907) writing a summary of my work on the theory of special relativity…, I also had to try to modify the Newtonian theory of gravitation such as to fit its laws into the theory. While attempts in this direction showed the practicality of this enterprise, they did not satisfy me because they would have had to be based upon unfounded physical hypotheses. At that moment I got the happiest thought of my life[1] in the following form:
>
> In an example worth considering, the gravitational field has a relative existence only in a manner similar to the electric field generated by magneto-electric induction [Faraday's experiment]. Because for an observer in free-fall from the roof of a house there is during the fall – at least in his immediate vicinity – no gravitational field. Namely, if the observer lets go of any bodies, they remain relative to him, in a state of rest or uniform motion, independent of their special chemical or physical nature. (Of course, this consideration ignores the air resistance.) The observer, therefore, is justified in interpreting his state as being 'at rest.' [Emphasis in original][2]

This beautiful and simple idea is reasonably easy to explain, especially with the visual aid of an elevator. For this Einstein may have drawn on his own experience. With the invention of the electric motor in mid-century, in 1857 the first electric elevator appeared, and it was a practical prerequisite for a major innovation in architecture that still dominates cities today – the skyscraper. It is not known how extensive was Einstein's exposure as a child to the rise and fall of an elevator, but he used the image as an adult in his exposition of the theory.[3] While riding in an elevator, one's

---

[1] "…der Glücklichste Gedanke meines Lebens…": that is, the happiest, or luckiest, or most fortunate, or most successful thought of my life. *Einstein Papers*, Vol. 7, Doc. 31, p. 265.

[2] *Einstein Papers*, Vol. 7, Doc. 31, pp. 135–136 ET. In the 1922 Kyoto lecture, which must be used with caution (as explained before), he attributes the genesis of this idea to a specific moment while working in the patent office in Bern. "Suddenly an idea dawned on me: 'If a man falls freely, he should not fell his weight himself.' I felt startled at once. This simple thought left me with a deep impression indeed." Quote in Abiko [1], p. 15.

[3] In the popular account he uses this image without calling it an elevator. Einstein [49] [1917], pp. 66–70.

D.R. Topper, *How Einstein Created Relativity out of Physics and Astronomy*, Astrophysics and Space Science Library 394, DOI 10.1007/978-1-4614-4782-5_12, © Springer Science+Business Media New York 2013

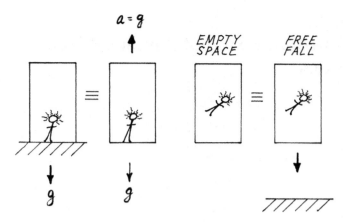

Fig. 12.1 Einstein's 1907 thought experiment on the identity of gravity and acceleration. There is no experiment the person in the elevator can perform in order to distinguish between gravity and acceleration. The two cases on the *left* and the two on the *right* are each experientially identical

apparent weight increases and decreases as one moves up and down, respectively. This is due to a change of state, using Galilean terminology, whereby the force of gravity is, in a sense, increased or decreased.[4]

This experience was then extrapolated to develop his theory of gravity. Consider Fig. 12.1: the image is of Albert confined to a closed space, under different physical circumstances. First consider the two cases on the left. If the elevator is at rest on Earth, he feels the force of gravity with a power of **g** pulling downward. If, on the contrary, the elevator is in empty space but moving upward and accelerated at the same rate, he also feels a force downward of **g**. Within the elevator Albert experiences the same force as on Earth, although this is an inertial force due to his being in a non-inertial system. Within his world, since he cannot see outside his elevator, he is not able to distinguish between the two different sources of force, one gravitational and the other inertial. The symbol ≡ indicates an equivalence between the two cases. Put in Einsteinean experiential terms: there is no experiment he can perform, say by dropping an apple or a rock, to tell the difference between the two cases.

Second, consider the two cases on the right. Now Einstein is floating in empty space, where he experiences no gravitational force. Here Albert floats within the elevator and anything with him floats along too, say the apple or rock. If, however, we transport the elevator back to Earth and drop it from a great distance, he and the elevator are in free fall (just as if, while on Earth, an elevator cable breaks). Under this condition he and the elevator fall all at the same rate (along with the apple or rock) – because of Galileo's discovery over 400 years ago that bodies fall independently of their weight – and his experience is identical to floating in empty space,

---

[4] Einstein once joked with a *New York Times* reporter, that he saw a man fall from a roof and ran up to him asking what it felt like. Others picked–up on this and another myth was created. Clark [26], p. 303.

despite the fact that he is near the Earth and falling within a gravitational field. As before, this identification of experiences is symbolized with the≡between floating in empty space and being in free fall on the Earth. Put again in Einsteinean experiential terms: there is no experiment he can perform to tell the difference between the two cases.

This last free-fall case was the key insight – the happiest though of his life – for Einstein sometime in 1907. He was so enamored by this that he underlined it in the 1920 document. As he said, this case of free-fall is experienced as if gravity is turned off, similar to the fact that electricity can be turned on and off.

Now to the second part of the document, the remainder of which should be clear, partially from the previous analysis and also with some further scrutiny:

> The extremely strange and confirmed experience that all bodies in the same gravitational field fall with the same acceleration [,] immediately attains, through this idea, a deep physical meaning. Because if there were just one single thing to fall in a gravitational field in a manner different from all others, the observer could recognize from it that he is in a gravitational field and that he is falling. But if such a thing does not exist – as experience had shown with high precision – then there is no objective reason for the observer to consider himself as falling in a gravitational field. To the contrary, he has every right to consider himself in a state of rest and his vicinity as free of fields as far as gravitation is concerned.
>
> The experimental fact that the acceleration in free-fall is independent of the material, therefore, is a powerful argument in favor of expanding the postulate of relativity to coordinate systems moving nonuniformly relative to each other.[5]

There are two essential points in this passage: first, the reference to experimental evidence for Galileo's deduction; and second, the inference from this fact to extending the principle of relativity to non-inertial systems. On the first point: historically there are a number of experiments relevant to Einstein's assertion, and he seems to be alluding to at least two of them. In chronological order they are: Galileo's law of falling bodies, discussed above, and Newton's experiment on gravitational and inertial mass.

Newton employed Galileo's argument for bodies falling independently of their weight. He also replaced the relative concept of weight with the absolute concept of mass. But, of importance to this Chapter is that there were fundamentally two concepts of mass in Newton's physics. The terminology used here is that of Einstein, not Newton, but the concepts are the same. Consider the case of a body (say, the 12 lb rock mentioned before) being attracted by the Earth; for this situation the gravitational mass of the rock would be used in any calculations. Yet place the same rock on a smooth table, perhaps oiled to minimize friction, and begin pushing it; the rock resists this force, and the resisting force can be measured. This resistance is due to inertia – gravity plays no role on a horizontal table – and it is called, appropriately, inertial mass. Thus there are two concept of mass, gravitational and inertial. As we saw in Galileo's deduction on falling bodies, his insight into how bodies accelerate as they fall implied (in our new terminology) a proportionality between gravitational and inertial mass (Fig. 11.2 was a visual way of saying the same thing). So, the two masses seem to be identical, but are they?

---

[5] *Einstein Papers*, Vol. 7, Doc. 31, pp. 135–36 ET.

Their identification was part of the theoretical structure of Newton's physics, but he wanted to test it empirically. He did so by using pendulums (the bobs being the masses). In an oscillating pendulum both concepts of the masses come into play: the gravitational mass, since the bob is falling under gravity, and inertial mass since the bob is moving back and forth in a curved arc. Newton made two identical pendulums with 11-ft cords and with bobs made out of thin hollow shells in which he put nine different substances in pairs: gold, silver, lead, glass, sand, common salt, water, wood, and wheat. Always putting the same weight of each substance into the two bobs, he then set them in motion at the same time looking for a difference in their periods of oscillation. With the weights being the same, the shapes being the same, and even the air resistance being the same for both pendulums, then any difference in oscillation could only be due to a difference in inertial mass due to different substances being in each pair of bobs. None was found: all pairs of oscillation were identical. Newton thus concluded that the identification of inertial and gravitational mass is an empirical fact. He estimated the accuracy: "In these experiments, in bodies of the same weight [that is, gravitational mass], a difference of matter [that is, inertial mass] that would be even less than a thousandth part of the whole could have been clearly noticed."[6] When Einstein, in the 1920 quotation, used the phrase "as experience had shown with high precision" in referring to objects falling independently of their weight, he was likely citing Newton's experiment (accurate to 1/1,000) along with Galileo's.

There is, however, a third possibility: the Hungarian physicist, Baron Roland von Eötvös in 1888 began performing a series of extremely delicate experiments looking for such a difference in mass using a version of the torsion balance that Cavendish and Coulomb had used to measure gravitational and electrical forces, respectively. He too found no difference in the gravitational and inertial masses, and Eötvös estimated that his result was accurate to one in a billion. His early experiments were published in 1890 and translated into German in 1891. In the 1920 document, Einstein did mention Eötvös' experiments, about which he used the term "extraordinary[7] precision," but this came two pages *after* reminiscing on the 1907 insight, the implication being that he was not aware of the work of Eötvös in 1907. In 1933, in a lecture in Glasgow, Einstein recalled the genesis of general relativity, and specifically the identification of inertial and gravitational mass, and said: "I had no serious doubts about its strict validity even without knowing the results of the admirable experiments of Eötvös, which – if my memory is right – I only came to know later."[8] Thus I assume the "high[9] precision" reference in the 1920 document was only to Newton's experiments. The earliest reference I can find is a 1918 letter to Eötvös on non-scientific matters, in

---

[6] Newton [151] [1726], Book 3, Proposition 6, Theorem 6, p. 807.

[7] *Ausserordentlicher*, in German; *Einstein Papers*, Vol. 7, Doc. 31.

[8] Einstein, "Notes on the Origin of the General Theory of Relativity," reprinted in Einstein [47], pp. 287.

[9] *Grossser*, in German.

which Einstein mentioned in passing that he wished to express his "gratitude for the advancement that our knowledge of the identity of gravitational and inertial mass has made through your investigations."[10] This explains the 1920 reference to Eötvös.

In short, Einstein concluded that free-fall in a gravitational field is identical to floating in empty space, and, it appeared, the relativity principle could be extended to noninertial systems. Special (inertial) relativity gave way to general relativity, such that all motion (inertial and noninertial) was relative. At least, that was the goal.

This exceptional reminiscence from 1920 is so important and significant in Einstein's work that it is being quoted again, so you may read it through in one swoop, without my commentary interfering with your train of thought. Here it is in one piece:

> When I was busy (in 1907) writing a summary of my work on the theory of special relativity…, I also had to try to modify the Newtonian theory of gravitation such as to fit its laws into the theory. While attempts in this direction showed the practicality of this enterprise, they did not satisfy me because they would have had to be based upon unfounded physical hypotheses. At that moment I got the happiest [or luckiest, or most fortunate, or most successful] thought of my life in the following form:
> In an example worth considering, the gravitational field has a relative existence only in a manner similar to the electric field generated by magneto-electric induction. <u>Because for an observer in free-fall from the roof of a house there is during the fall</u> – at least in his immediate vicinity – <u>no gravitational field</u>. Namely, if the observer lets go of any bodies, they remain relative to him, in a state of rest or uniform motion, independent of their special chemical or physical nature. (Of course, this consideration ignores the air resistance.) The observer, therefore, is justified in interpreting his state as being "at rest."
> The extremely strange and confirmed experience that all bodies in the same gravitational field fall with the same acceleration [,] immediately attains, through this idea, a deep physical meaning. Because if there were just one single thing to fall in a gravitational field in a manner different from all others, the observer could recognize from it that he is in a gradational field and that he is falling. But if such a thing does not exist – as experience had shown with high precision – then there is no objective reason for the observer to consider himself as falling in a gravitational field. To the contrary, he has every right to consider himself in a state of rest and his vicinity as free of fields as far as gravitation is concerned.
> The experimental fact that the acceleration in free-fall is independent of the material, therefore, is a powerful argument in favor of expanding the postulate of relativity to coordinate systems moving nonuniformly relative to each other.

All motion came under the umbrella of relativity. From this point of view, gravity, as an occult force, could be eliminated by substituting a non-inertial system, thanks to the equivalence between inertial and gravitational mass.

***

The 1920 document is one source of evidence that Einstein began thinking of extending relativity to include acceleration in 1907. We also know that he carried the idea even further at the time, for the last short section of the 1907 article in Stark's journal contained a preliminary extension of the idea; in fact, he made two predictions from it. In this important section, titled, "Principle of Relativity and

---

[10] *Einstein Papers*, Vol. 8, Doc. 450.

Gravitation"[11] he showed that there was also a time dilation for gravity; that is, clocks run more slowly the stronger the gravitational field.

The idea that gravity slows time may be grasped intuitively from thinking of general relativity an as extension of special theory. Since the passage of time of a moving clock slows-down in proportion to its constant speed, then the speeding-up of an accelerating clock would also (or further) slow-down the passage of time, and in proportion to the acceleration; in addition, since acceleration is equivalent to gravity, then a clock in a gravitational field would "tick" more slowly in proportion to the strength of the field.[12] Put simply: a clock at sea level runs slower than a clock at the top of a mountain. Unfortunately the effect was too small to detect using clocks at the time. Nonetheless, this was the first prediction from the 1907 essay.

Next, in a short paragraph, based on the gravitational time dilation, Einstein made a second prediction: that light from the Sun should be very slightly shifted toward the red part of the spectrum. If this is not obvious – and it was not the first time I read it – then here is an explanation that should make it clear. Consider a chemical element at the surface of the Sun, an element we can indentify by its spectrum. That element is made-up of atoms which vibrate, which is the source of the spectrum of light. A vibrating atom is in essence a clock; indeed, vibrating atoms were later used to make atomic clocks, for they vibrate in micro-fractions of a second at constant rates (or frequencies). It follows, according to the time dilation from gravity, that an atom on the Sun should vibrate (that is, mark time) at a slower rate than the same atom on the Earth, since the Sun's mass and thus its gravity is much stronger than the Earth's. A slower vibrating atom has a longer wavelength[13] and this would be exhibited as a shift toward the longer (red) end of the spectrum. Einstein thus predicted that comparing the spectrum of an element on the Sun with the same element on Earth would show a solar redshift due to the stronger gravity of the Sun. This later became known as the gravitational redshift.[14] As seen later in this Chapter, these predictions were two of four key predictions of what became his general theory of relativity. Both were deductions following the postulate of the equivalence between inertial and gravitational mass, whose source was the thought experiment involving free fall.

There was one more source of the equivalence postulate, and so another route to the relativity of acceleration that we must explore before picking-up and following Einstein's path toward the general theory. It began when he read a book by the Austrian physicist-philosopher, Ernst Mach,[15] mentioned briefly before. But I save that story for next Chapter.

***

---

[11] *Einstein Papers*, Vol. 2, Doc. 47, pp. 301–311 ET.

[12] The quantity in general relativity will be given later.

[13] Wavelength (a distance) times frequency (a number/time)=the speed of light (c), which is a constant. If the frequency decreases then the wavelength increases.

[14] This phenomenon should not be confused with another redshift associated with the Doppler principle and the expanding model of the universe, an important topic in Part IV.

[15] (1838–1916).

At this juncture in a book weighted heavily on Einstein's thinking process, it is an auspicious to time pause briefly to look at a conceptual (or psychological) topic, since we have methodically mulled-over his two famous thought experiments. I wish now to draw attention to the visual or almost pictorial aspect of these imaginary experiments. Both may be conceived, as he did – and, indeed as I have drawn – namely, pictorially. Whether riding a beam of light or experiencing forces in an elevator involved palpable mental imagery. Einstein, himself was aware of this mode of thinking. Listen to this statement on his thought process written in 1944:

> The words or language, as they are written or spoken, do not seem to play any role in my mechanism of thought. The psychical entities which seem to serve as elements in thought are certain signs and more or less <u>clear images</u> which can be 'voluntarily' reproduced and combined.... [T]his combinatory play seems to be the essential feature in [my] productive thought – [this happens] before there is any connection with logical construction in words or other kinds of signs....The above mentioned elements are, in my case, of <u>visual</u> and some of muscular type.[16]

This statement, although rather cryptic, provides a window, however cloudy, into Einstein's thinking, or at least his thinking about his thinking, with its emphasis on the role of imagery and visualizing.

A few years later, writing his autobiography,[17] he raised the same topic in a section that began with the question: "What, precisely, is 'thinking'?" Again, the explanation was less than lucid, but it did make explicit that "pictures"[18] played a key role in the emergence of concepts, and this transition constitutes thinking. He went on to emphasize that this stage of "free play" with concepts took place "without [the] use of signs (words)" and further that this (possibly unconscious) mental activity was a form of "wonder." From this he segued into the two "wonders" mentioned before in this book: the behavior of a compass and the "holy" geometry of Euclid.[19]

The role of imagery and or visualizing in thought may be more plainly understood by a simple and lucid example (not, however, from Einstein) involving telling time.[20] Consider this problem: It is 1:50 pm, and you wish to know the time one half-hour later. There are at least two ways of solving this. You may add 30 min, and arrive at 2:20 arithmetically. The other obvious way is to make a mental picture of a (non-digital) clock, where the minute hand points to the 10, and mentally flip the hand 180° to the 4; this way you arrive at 2:20 pm. This latter method has been called visual thinking.[21] I'm convinced that it was something like this mental

---

[16] From a letter to Jacques Hadamard, quoted in Hadamard [85] [1949], pp. 142–143, emphasis is mine. The letter constitutes Appendix II to the book. Miller [144], p. 370, dates the letter as June 17, 1944.

[17] The draft was written in 1947.

[18] *Bilder*, in German.

[19] Einstein [51] [1949], pp. 7–8.

[20] It goes without saying that this has nothing to do with relativistic time.

[21] Arnheim [4].

processing that Einstein was articulating; the other, arithmetical approach, scientists also used.[22] Perhaps the work at the Patent office, daily dealing with illustrations of apparatuses, helped to hone his visual skills, which were already nascent in his early life – witness the image of riding a beam of light at age sixteen.[23]

Armed with this concept, we look at the role visual thinking played in Einstein's work later in life.

---

[22] Schweber [180] contrasts Einstein's pictorial approach with that of fellow physicist Robert Oppenheimer (who we will come across again later), which "was more analytic and formalistic," p. 15.

[23] For more on Einstein and imagery, see Miller [144], pp. 312–324 and 361–378.

Chapter 13
Enter, Mach's Principle; or, Seduced by an Idea

Sometime during Einstein's student years, his friend Besso introduced him to the widely-read book *The Science of Mechanics*, by Mach.[1] In perhaps the most famous section of the book, Mach put forward a critique of Newton's concept of absolute motion and the corresponding idea of absolute space. Einstein was enamored by this argument and pondered it for many years.[2] The argument from Newton, however, first must be understood before considering Mach's challenge. So we return to Newton and yet another famous thought experiment – this one is called Newton's bucket experiment.

Galileo's relativity of motion had a profound impact on Newton. I think it gnawed at the core of what he believed was the goal of science (what he called natural philosophy): to find God in nature. Since God is the absolute Absolute, then finding absolutes in nature was a step toward finding Him. Newton said as much in the *Principia* at the end of the introductory section containing what he called "Definitions," and which contained the bucket experiment. He said that the rest of the book was a "fuller explanation" of how to determine absolute motions from their causes and effects, and, conversely, to determine the causes and effects from the motions, relative or absolute. In this logical loop, a key part was the deduction of absolute motion; indeed, listen to the last sentence of the Definitions section: "For this [that is, finding absolute motion] was the purpose for which I composed the following treatise."[3] To me it says: 'I wrote the Principia to find God in nature.'

---

[1] *Die Mechanik in Inrer Entwicklung, Historisch-Kritisch Dargestellet*, the first edition published in 1883. Available in English as *The Science of Mechanics: A Critical and Historical Account of Its Development*. See Mach [134]. Notice how the title of Einstein's important 1920 document, "Fundamental Ideas and Methods of the Theory of Relativity, Presented in Their Development," discussed above, echoes the title of Mach's book. Stachel [192], pp. 262–263.

[2] The most comprehensive study of this topic is by Hoefer [95, 96]. It has been pointed out that there are at least ten different interpretations of what constitutes Mach's principle. See, for example, Bondi and Samuel [11], and the list in Barbour and Pfister [6], on p. 530. This more general problem is not of concern here, since we are only interested in how Einstein interpreted it.

[3] Newton [151] [1726], p. 415.

D.R. Topper, *How Einstein Created Relativity out of Physics and Astronomy*, Astrophysics and Space Science Library 394, DOI 10.1007/978-1-4614-4782-5_13,
© Springer Science+Business Media New York 2013

If true, Newton found God, so to speak, in a bucket of water, among other places. Not any bucket, however, but a rotating one. Let's see how.

His thought experiment began with a bucket filled with water and hanging from a long cord. If the bucket is at rest then the surface of the water is still. Now wind up the bucket so as to knot the cord, and then release it. As the bucket begins spinning the surface of the water forms an arc, rising at the edge; as the rotation increases, more water is forced toward the edge and eventually spills out of the bucket. The cause of this is what we call centrifugal force,[4] a force that appears when something rotates.

To Newton the difference between the phenomena associated with the bucket at rest (with the water being calm) and the bucket in motion (with the water spilling-out) was a clear-cut example of an experience that distinguishes between relative motion and absolute motion. To him it would be absurd to say that the spilling bucket is at rest, for how, otherwise, could one explain the spilling? It was a convincing argument for him and for many others too. Consequently, Newton's viewpoint prevailed for a long time.[5] I recall being completely swayed the first time I read it. Yet Mach made a case that the spinning bucket may be considered to be rest.

Before confronting his argument from the late-nineteenth century, we need first to pause in mid-century for an important experiment that is crucial to this story. In 1851 in the Rotunda of the Pantheon in Paris, the French physicist, Leon Foucault, demonstrated something he discovered originally in his laboratory about the motion of a short pendulum. Performing it on a large scale using a 100-ft pendulum in the Rotunda, he showed that the plane of the oscillating pendulum was not fixed, but that it slowly rotated – its speed, like a clock, being a function of the rotation of the Earth. A simple way to understand this is to picture a pendulum at the North Pole. As it oscillates back and forth with respect to the stars, the Earth rotates on its axis. From the point of view of someone at rest on the Earth and watching the pendulum, its plane rotates such that in one day it makes one rotation. The period, however, changes at different latitudes on Earth; at the extreme, a pendulum on the equator would have no rotation. At Paris, the period is less than a day based on a calculation using the latitude of that location. Foucault's discovering was interpreted as confirming Newton's conception of absolute motion, and even absolute space. The oscillating pendulum is clearly fixed with respect to the space in which the Earth is rotating; that is, the pendulum is not rotating along with the Earth. This means that the pendulum's plane is fixed in space, in an absolute space. Examples of so-called Foucault pendulums are found today in public places throughout the world, and are interpreted as proving the rotation of the Earth on its axis, just as stellar parallax,

---

[4]Centrifugal, is from the Latin words for a force away from the center; it was coined by the Dutch mathematician-physicist Christiaan Huygens, a contemporary of Newton. Incidentally, Newton subsequently coined the term, centripetal force, for a force toward the center.

[5]I am avoiding the challenges from the German mathematician-philosopher Gottfried Leibniz and others, since it is not relevant here, and the topic is far beyond the scope of this book. In some ways, Mach later picked up where Leibniz left off.

over a decade earlier, proved the motion of the Earth around the Sun. (On the latter topic, you may jump ahead to Fig. 20.1, where the proof is discussed.)

***

In a significant way Foucault's experiment harkens back to a splendid little experiment that Galileo performed. This, in turn, displays a connection with the concept of inertia, which we will see was important to Mach and Einstein too. In his experiment Galileo floated a ball in a bowl of water, holding the bowl carefully so that the ball remained in the center. Next he slowly and carefully rotated the bowl in a circle with respect to the room preventing the ball from touching the side. It should come as a surprise that although the bowl and the water were moving in a circle, the ball itself remained fixed with respect to the room. It did not rotate with the water.[6] This was one of Galileo's key examples of inertia in action: the ball remained at rest, in a state of rest, and resisted the motion of the encompassing water. Just as there is an inertial resistance of matter to a direct linear push or pull, in this case the inertia of matter resisted a rotation or a torque.[7] For Galileo the context of this ball remaining in a fixed position with respect to the room was applied to the physics of astronomy; specifically, to explain how the Earth as it moves around the Sun remained in a fixed tilted position with respect to the stars. He therefore made a conceptual leap from the fixed ball (in the room) to the fixed tilt of the Earth (in the universe). This also shows why it is mentioned it here: for we now make the analogy between Galileo's ball in a bowl of water and Foucault's pendulum bob as examples of bodies of matter exhibiting resistance to motion. Foucault's pendulum remained fixed with respect to the stars, as the Earth turned underneath; and this, like Galileo's ball, was an example of inertial resistance of a change of rotational state. Another way of expressing this fixed position with respect to the stars was to speak in terms of the absolute space of Newton. That indeed was how Foucault's pendulum was widely interpreted, as confirming Newton's absolute space. Now, back to Mach.

***

As a philosopher, Mach wanted to base scientific theory entirely upon empirical knowledge, purged of what he called "metaphysical obscurities."[8] In the case of motion, what is observed is only relative motion. A ship is moving with respect to the shore, but within the enclosed cabin everything happened as if the ship were at rest. Thus, linear motion is relative motion, and importantly this motion is within relative space. Absolute space was a metaphysical notion or a construct of the mind according to Mach. If all space is relative space then a rotation within that space

---

[6] Actually the ball does minimally rotate, due to the friction of the water, but it is clear that without friction there would be no rotational motion, just was without air resistance all falling bodies would fall at the same rate. I recommend the reader perform this simple experiment; it is quite astonishing.

[7] Galileo [71] [1632], pp. 389–399.

[8] Mach [134] [1883, preface to the first edition], p. xxii.

must also be relative motion. So how are we to explain the motion of the plane of the Foucault pendulum?

Mach's explanation I find cryptic. I have, however, come-up with a way of clarifying it, at least for myself, and I hope for the reader too. A clear-cut way to approach this question is with a sort of historical thought experiment where I put myself in the late-Middle Ages. I am a scholar steeped in the geocentric universe and living near Stonehenge (around 50° north latitude, which is about where I really am now in Winnipeg, Canada). While working with a pendulum, I discover the rotating plane. Timing the motion I find that it is almost a day. How do I interpret this discovery? Within my cosmic model, a day is the time of rotation of the celestial sphere carrying along the Sun and Moon. Moreover, within this model the celestial sphere has a pole set at an angle, for me, 50° from the horizon, which is marked by a star we appropriately name Polaris (the pole star). With a bit of mental gymnastics, I realize that the departure of the period of the pendulum's plane from an exact day corresponds to my distance from being directly under the North Star. Recall that European scientists in the Middle Ages knew the Earth was a sphere and that the stars appeared in different positions at different places on Earth. From this I surmise that the pendulum's plane of rotation would an exact day if I were at the North Pole directly under Polaris, with all the stars rotating around.

Having arrived at this conclusion, the question of causality naturally arises: What is pulling the pendulum bob around in a circle? The only answer, it seems, is the existence of some power acting-at-a-distance between the stars and pendulum bob, since the plane of the pendulum bob moves in a circular motion along with the stellar sphere around the Earth, while the Earth is motionless at the center of the cosmos. This appears to be another case of an occult power, such as the magnetic attraction between a compass and the North Star. If I am an Aristotelian scholar, I may be leery of accepting the existence of an occult power, but without an alternative explanation of the phenomenon.

Let us now scoot forward over half a millennium to the nineteenth century. Obviously Mach certainly did not believe in the geocentric model. Indeed late-nineteenth century scientists drew on the discovery of stellar parallax and the Foucault pendulum as empirical evidence for heliocentrism and the rotating Earth, respectively. But the heliocentric explanation of the Foucault pendulum was based on the notion of absolute motion and absolute space, where the pendulum is at rest with respect to this space, with the Earth moving under it. This we perceive as the rotation of the plane of the pendulum bob.

Mach, however, had no patience with the concept of absolute space, which he even called "monstrous."[9] How then to explain the phenomenon? There seemed to be only one answer, the attraction of the stars to which the pendulum remains fixed, just as my imaginary medieval scholar suggested from a geocentric viewpoint. Living in a post-Newtonian world, Mach was less reticent to shy away from powers acting-at-a-distance. After all, gravity was an inverse-squared force extending throughout space

---

[9]Mach [134] [1912, from the Preface to seventh edition], p. xxviii.

between all bodies of matter. The gravity of the fixed stars therefore accounted for the fixed position of the pendulum's swing, not the metaphysical concept of absolute space.

Moreover, there was a bonus: Mach not only eliminated absolute space but he also provided an explanation for inertia, a causal explanation. This may require further explanation, for this deduction is not obvious. Indeed, I think Mach's idea is best explained by going back to Galileo. As seen, he made progress in the science of motion by shifting the role of the scientist's questioning of nature from asking "why" to asking "how." Inertia was a concept based on how bodies behave under diverse conditions. Thus the ball in a bowl of water remained in a fixed position, even if the water and bowl were rotated in a circle around the ball. It remained at rest, in a state of rest; it resisted the rotation of the water around it. That was how it behaves. The behavior was later called inertia. If we now ask the question as to why the ball behaved this way, an answer may go something like this: the ball was in a fixed position, not only with respect to the immediate room, or to the local world, or to the Earth, but ultimately to the stars, fixed in space; they are the cause, not the "monstrous" concept of absolute space. What we call inertia in this case, is really the attractive power of the stars on the ball.

Mach probably was not aware of Galileo's ball in a bowl example, but in essence my previous analysis of it is how Mach explained the motion of the pendulum. Mach's explanation for the plane of the bob's fixed motion was the attractive power of the stars; this meant that the explanation of inertia (the why, in addition to the how) was the fixed stars. Indeed, consider a single mass in an empty universe: there is no way of deciding if it is moving, neither linearly nor rotationally. This also means not only that all motion is relative for this single mass, but there is no centrifugal force and no inertia, without the attraction of the stars.

This brings us back to Newton, for the same analysis applies to his bucket and therefore to the cause of centrifugal forces. Mach's analysis implied that rotational motion, like linear motion, was relative motion, since there was only relative space. Therefore the rotating bucket may be considered as being at rest; but how then to explain the splashing water? Something must be pulling the water out of the bucket. That something can only be a force fixed with respect to the bucket; namely, the attraction of the distant stars. Although the immediate case of the splashing water is attributed to centrifugal forces, the ultimate cause is the attraction of the stars.

All this is a very seductive idea, allowing the generalizing of relativity to include both linear and rotational motion; or, the same thing said from another viewpoint, it integrated both constant speed and acceleration under the relativity principle; or, yet, within a further framework, the relativity principle was applied to both inertial and non-inertial systems. However expressed, it came with the bonus of explaining for the first time the source or cause (or why) of inertia. I believe this apparent causal explanation of inertia was its key attraction. Einstein certainly was captivated by Mach's principle, as he later called it.[10] In a (1912) paper with an interesting

---

[10]Einstein first used the term "Mach's principle" in 1918. *Einstein Papers*, Vol. 7, Doc. 4, "On the Foundations of the General Theory of Relativity." But the idea appears much earlier in his work.

question for a title, "Is there a Gravitational Effect which is Analogous to Electrodynamic Induction?," he deduced a case where it seemed "that the entire [his emphasis] inertia of a mass point is an effect of the presence of all other masses, which is based on a kind of interaction with the latter."[11] Thus, to repeat, for a single mass in an empty universe, all motion is relative and there is no absolute rotation, no centrifugal force, and no inertia.

Around the same time Einstein wrote to Mach saying that if his (Einstein's) idea about the equivalence of relativity and gravity is correct then "it follows of necessity that inertia [his emphasis] has its origin in some kind of interaction of the bodies, exactly in accordance with your [i.e., Mach's] argument about Newton's bucket experiment."[12] In sum, this was another route to the relativity of acceleration, along with the thought experiment of the equivalence of falling in a gravitational field and floating in empty space.

Out of this fertile mixture of concepts and experiments – from Galileo and Newton to Foucault and Mach – and after years of effort, Einstein produced the general theory of relativity, what many historians believe will probably be the last revolutionary theory in science formulated by essentially one person.

---

[11] *Einstein Papers*, Vol. 4, Doc. 7.
[12] Letter to Mach, 25 June 1913, *Einstein Papers*, Vol. 5, Doc. 448.

Chapter 14
Einstein's Epic Intellectual Journey: 1907 to 1915

There are several cases of what I call epic journeys in the history of science. Not Ulysses-like journeys on land and sea, but journeys of the mind. Kepler's wrestling with data of the planet Mars, from which he deduced its elliptical orbit, was a solitary achievement of momentous consequent. In October 1601, upon the death of the observational astronomer extraordinaire, the Dane, Tycho Brahe, Kepler inherited Tycho's job and the most up-to-date and detailed observational data of Mars ever. Probably early in 1602 he began his quest to calculate the orbit of Mars from Tycho's data. Kepler wagered that he would finish the task in a few weeks; it took him over three years, during the course of which he called his struggle his "war with Mars." He finally deduced the elliptical path of the planet around Easter 1605 – a truly epic journey, the details of which are still being scrutinized by historians poring over Kepler's notes and writings. The same is true for Einstein's struggle with formulating a gravitational theory of relativity, which too was nearly solitary over his eight-year slog.[1]

The details, regrettably, are buried within dense tensor calculus equations that are penetrable only to mathematical physicists.[2] Fortunately, the physical concepts engaged in the theory can be visualized with the aid of some analog thinking and are plainly explained with diagrams. So, here goes...

---

[1] The intellectual journey from 1907 to 1915 was not necessarily a continuous process. There is a publishing a gap on the topic of gravity from late-1907 to mid-1911, with little discussion too in his correspondence. There are several reasons for this: he was publishing extensively in quantum physics over those years; he was very busy with teaching and administrative duties for the first time in his life; and he may have been "stuck" on how to turn the 1907 thought experiment into a theory of gravity. However, in the 1933 Glasgow lecture Einstein said that the problem of gravity "kept me busy from 1908 to 1911" (Einstein [47], p. 287), although this does not contradict the possibility that he was "stuck." See Pais [162], pp. 187–190.

[2] Pais [162], Part IV; for this story, along with more recent citations of the scholarship of others, see Van Dongen [205], Chap. 1. Also, Jungnickel and McCormmach [111], pp. 321–347, especially for the increasing role of mathematic in theoretical physics.

D.R. Topper, *How Einstein Created Relativity out of Physics and Astronomy*, Astrophysics and Space Science Library 394, DOI 10.1007/978-1-4614-4782-5_14,
© Springer Science+Business Media New York 2013

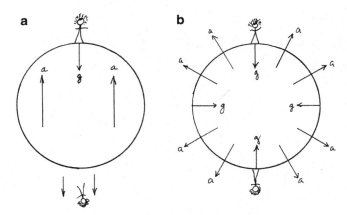

Fig. 14.1 Diagrams showing two unsuccessful ways of applying Einstein's equivalence argument to explain gravity on the Earth

The key insight and assumption of general relativity is, as seen in Chapter 12, the thought experiment that the experience of being in a gravitational field is identical to being in an accelerating frame of reference. This, as in Fig. 12.1, is easily visualized. The next problem is to turn this into a theory of gravity. One way it will not work is by accelerating the Earth in a specific direction (as in Fig. 14.1a), for although Albert will experience a gravitational force, the woman on the opposite side will be accelerated into space. Nor can we explode the Earth (Fig. 14.1b). How then can the accelerations of two falling bodies on opposite sides of the Earth, 180° apart, be due at the same time to being in a non-inertial system?

The answer lies in the space around and between them, a specific concept of space. This space is not a passive receptacle for matter; it may be empty but it is not nothing. This space has properties such as elasticity, and as such it can bend, warp, or curve around things, such as matter. In essence, matter bends space and the bending of space is what we experience as a gravitation force; so the distortion of space around matter is what we call the gravitational field. Faraday's field was space itself! But I am jumping ahead.

Einstein specifically found the answer, perhaps initially to his chagrin, in Minkowski's extension of space into the fourth dimension, an idea that he said mystified rather than clarified his theory. In four-dimensional space-time, gravity and inertia may be equivalent, if another extension is granted: the four-dimensional space is non-Euclidean. Such a space is pliable; that is, the space may be bent (or warped or curved, depending upon your adjective of choice), so that the gravitational field is due to this curving of space around a mass rather than a spooky occult power acting-at-a-distance. To understand this, let us review some more basics about geometry.

Euclid's "holy" geometry book began with five postulates, statements taken for granted as being true by inspection. Since around 300 BCE when the treatise was written, the first four were indeed taken for granted, but over the centuries the fifth postulate was a source of some discussion. One way of expressing Euclid's fifth

Fig. 14.2 Euclid's fifth
postulate. There is one and
only one line **L'** through **P**,
in the plane of **P** and **L**,
which is parallel to **L**

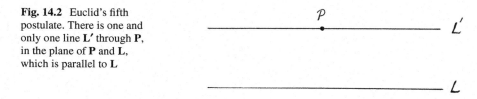

postulate is this (Fig. 14.2): given a straight line **L** and a point **P** not on the line, there
is one and only one line **L'** through **P**, in the plane of **P** and **L**, which is parallel to **L**.
Occasionally over almost two millennia the question was posed: can the fifth postulate be deduced from the first four? All attempts to do so were unsuccessful. In the
nineteenth century, four mathematicians asked a different question: if we only use
the first four postulates, what sort of geometry would we get? Most mathematicians
over the ages, if asked this question would probably have said something like this: an
effort to deduce a logically consistent geometry from only the first four postulates of
Euclid is an exercise in futility, since there is only one geometry; the task would at
least result in conflicting and contradictory corollaries, theorems, and other deductions. What the four[3] found, however, was, on the contrary, that logically consistent
deductions were arrived at, although different from those found by Euclid. In Euclid's
geometry, for example, the sum of the angles of a triangle is 180°. They found that
the sum could be more or less than 180° but not both. Regarding the fifth postulate
itself (Fig. 14.2): they found that there were either no lines through **P** parallel to **L**,
or an infinite number lines parallel to **L**, but also not both. Although, as expected,
other mathematicians initially balked at the notion of other geometries, by the
late-nineteenth century these non-Euclidean geometries found their way into the
corpus of mathematics, albeit as geometries of imagination or fantasy. After all, ever
since Newton, the space of the universe was defined as Euclidean, and this was
deemed to be the only real space.

This brings us back to Einstein, for this non-Euclidean idea was the second part
added to the four-dimensional framework of Minkowski. This is not obvious, so let
me explain. Among popular writers on relativity there is sometimes confusion and
conflation about extra dimensions and non-Euclidean geometry. They are distinct:
Minkowski's space is Euclidean; even with four-dimensions, it is still Euclidean. Or
even five-dimensions. Indeed, there are further dimensional spaces, going to infinite
dimensions, which nevertheless remain Euclidean. Extra dimensions are independent of being Euclidean or non-Euclidean. It is worthwhile going over this with
some simple cases before we return to Einstein and what becomes the first application of non-Euclidean geometry to the physical world.

A straight line is one-dimensional. This page (or the screen you are reading) is
two-dimensional. The room you are in is three-dimensional. A fourth- and more-
dimensional world can be conceived but not seen. All these are Euclidian. Return to

---

[3]They were: Carl Friedrich Gauss and his pupil Bernhard Riemann in Germany, Janos Bólyai in
Hungary, and Nikolai I. Lobachevski in Russia.

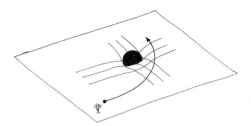

Fig. 14.3 Non-Euclidean space. Gravity is explained by the warping of space into an extra dimension; for us gravity is explained by space bending into the fourth-dimension. The illustration is an analog, showing how a 2-D person experiences an action-at-distance force for the warping of space into the third dimension

the case of the bug crawling on a straight wire (Fig. 8.1b). Now, put a kink in the straight wire, such that we distort the molecular structure of the wire, and the bug will notice a pull or push as it crawls over the kink. Next replace the two-dimensional paper page by a rubber sheet which is distorted by a mass into the third-dimension. This distortion of space is non-Euclidean, like the kink in the wire. Let's replace the bug with a two-dimensional man, call him Albert. This simple case is illustrated in Fig. 14.3, where Albert throws an object (a small mass) near a larger mass and perceives the object as being attracted to the mass. Such a warping of the sheet/space around the large mass would be detected by Albert as an attraction to the mass, like the push or pull on the bug in the kink of the wire. It should be understood that Albert, as a two-dimensional person, sees the mass as a disc, and the object he throws is also a disc. What is important is this example is that the apparent attraction between masses (two discs, for Albert) is due to the curvature of space around the larger mass into the third-dimension. The analog of this for us is that matter bends three-dimensional space in its vicinity into a fourth-dimension, which we do not see but detect as an attractive force. The introduction of a four-dimensional non-Euclidean space into the physics of motion, which we may picture by way of our analog thinking about our two-dimensional person's experience with a three-dimensional distortion of his space, was the culmination of Einstein's epic journey. This is, at least, the visualizable component of that journey.

Visually simple it was. The mathematical formulation, however, was another, very convoluted, effort. This is because of the interaction between the space and matter. Not being flat, the space is distorted by all bodies of matter, and thus the space is variable at all points of space. The distorted space, which is gravity, also moves matter, but matter, in turn, distorts space. Besides, remember that what I am calling space is really space-time. Now try to express this in a mathematical language in four-dimensions: that is a formidable task, so much so that Einstein – yes, even Einstein – needed help. Recall his distain of mathematics beyond its usefulness in physics. So great was his rebuff that he often cut mathematics classes at the ETH. His friend Marcel Grossmann, thankfully, did not. Furthermore, he went on to a professorship in mathematics at the ETH, and was an expert in – thankfully, did you guess it? – non-Euclidean geometry. In 1912 Einstein moved from Prague back to a position at the ETH, auspiciously timed for Grossmann to come to his rescue for a third time. John Stachel quotes Einstein at one point saying: "Grossmann, you must help me, or else I'll go crazy!"[4] Could anyone be so lucky? As Einstein wrote about his mathematical difficulties at this time: "Compared with this [gravitational] problem, the original theory of relativity is child's play."[5] Accordingly, Marcel taught Albert how to manipulate the cumbersome tensor calculus equations required in order to formulate a gravitational force within a four-dimensional, non-Euclidean space.

Together they began the merging of the imaginary geometrical world and the physics of motion of masses. No longer confined to fantasy, the new branch of mathematics found a physical application. Around 1913 several key papers written on the road to general relativity had dual authors (Einstein and Grossmann), until approaching 1915 there was a single author. It was an (almost) solitary journey for Einstein.

Beyond just explaining tensor calculus in words, I wish to mention a formal aspect of the mathematics at one important point in the journey. Around 1913, when Einstein and Grossmann were developing the first tensor equations to express gravity, they did not assume that the equations were covariant. (Recall from the previous chapter that the Minkowski space-time distance was a covariant term; namely, having the same form in all inertial systems.) It was not until 1915 that Einstein introduced the covariant rule back into the mathematics and thus succeeded in formulating the general theory. He realized, in retrospect, his mistake. Later, when Einstein worked on the search for a unified field theory, the aesthetic nature of the covariant rule would be a guiding principle in both his calculations and his philosophical speculations on the nature of science.[6]

Another key episode occurred earlier in 1911 when Einstein predicted that light would be bent by a gravitational field.[7] The effect, if there was one, only applied to very large masses. (In essence: special relativity is about a world of objects traveling near the speed of light, with $E = mc^2$; and general relativity is about a world, later a universe, of very large masses. In both cases, these were worlds of extremes.) The closest large mass is the Sun and light beams from the stars are handy light-rays. The problem is that when the Sun is out the stars cannot be seen, and on a clear night filled with stars, the Sun in on the other side of the Earth. Except in one instance: the rare and local event of a solar eclipse. During a solar eclipse the Sun is overhead and the stars are visible for several minutes as the Moon passes in front of the Sun blocking the sunlight and darkening the sky. According to the prediction of general relativity (as shown in Fig. 14.4) the bent light from a star at point **A** should be perceived

---

[4] Stachel [192], p. 281, attributes the source of the quotation to a fellow mathematics professor at the ETH, who had also been a friend of Einstein when they were students. Quoted too in Pais [162], p. 212, who also includes the German, "*Grossmann, Du must mir helfen, sonst werd' ich verrückt!*"

[5] Letter of 29 October 1912 to Sommerfeld, in *Einstein Papers*, Vol. 5, Doc 421.

[6] Van Dongen [205], p. 120.

[7] "On the Influence of Gravitation on the Propagation of Light," *Einstein Papers*, Vol. 3, Doc. 23.

Fig. 14.4 Einstein's
prediction of the bending
of light by gravity. In this
case, the star's light is bent
around the Sun during a
solar eclipse. The
prediction is that the star,
really at **A**, should appear
at **B** during the eclipse

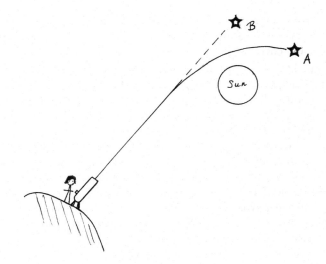

as a shift or deflection in the position of the star to point **B**. Consequently all stars in
the vicinity of the Sun at the time of the eclipse should exhibit such a deflection. By
comparing photographs of these stars during the eclipse with photographs of the
same area of the night sky without the Sun, the relative deflection of the stars should
be seen and measured. Therefore, "during a solar eclipse," he wrote, "it is possible
to compare this consequence of the theory with experience."[8]

A slightly amusing but potentially tragic story unfolded in August, 1914 when a
solar eclipse was occurring in Crimea and a German expedition of scientists traveled
to the place of the eclipse to photograph the stars around the Sun. Unfortunately this
was the outbreak of the First World War and they were captured by Russian soldiers
and almost sent to Russia as prisoners of war, but they convinced their captors that
they were scientists and were set free; their equipment, however, was confiscated and
not returned until after the war. Einstein, upon hearing of the unsuccessful experimental test of his theory, was disappointed. As we shall, it may have been fortunate
for Einstein that this eclipse experiment did not come off.

The last years of this intellectual journey were particularly trying for Einstein.
Not only due to the grueling effort required in wrestling with the tensor equations of
general relativity, but because of events in his life. In context, this intellectual journey was accompanied by a personal one: in the biographical parts of the previous
Chapters we saw Einstein and his family move from Bern to Zürich to Prague, and
back again to Zürich, with the offer in Berlin happening in the spring of 1913, and
finally the move coming the following spring. Shortly after he, Mileva, and their
children arrived in Berlin, Mileva left the city and returned to Zürich with the boys;
a possible divorce was discussed, along with an accompanying on and off estrange-
ment of Albert from his children. Close acquaintances of Albert and Mileva were

---

[8]*Einstein Papers*, Vol. 3, Doc. 23, p. 387 ET.

not surprised that they split. For many years their relationship was strained, with Albert being obsessed with his physics and devoting much less effort to Mileva and the family than she expected.

The move to Berlin also brought to fruition another irritant between them: for several years Einstein had gotten closer to a divorced cousin living in Berlin, Elsa Löwenthal,[9] who lived with her two daughters. Einstein knew her since childhood, when they often played together. During previous trips to scientific conferences and meetings in Berlin, he had visited his cousin and a bond developed that went beyond mere cousinly friendship. Apparently the physical presence of Elsa in Berlin was unbearable for Mileva, and so she departed with the boys. Einstein thus was alone in the summer of 1914, living in a bachelor apartment.

The trauma of the breakup of his marriage, and the loss of his children, was compounded in August with the start of the War and the accompanying excessive German nationalistic jingoism that he found offensive. He refused to sign a patriotic declaration in support of German militarism, a document directed to the outside world that was signed by the 93 intellectuals (scientists, artists, physicians, theologians, and others). On the contrary, he signed, along with only three other men, an alternative anti-nationalist manifesto.[10] He wrote to Ehrenfest, "The international catastrophe has imposed a heavy burden upon me as an internationalist."[11] Einstein's political views throughout this life were left of center, with him promoting in his non-scientific essays variations of socialism, the need for a world government, and a strong antipathy toward capitalism.[12]

His melancholy during the War, however, was tempered by his relationship with Elsa. Not long after Mileva and the boys left, he moved into an apartment in Elsa's building and was having meals with her. She gave him the physical attention he craved, the comfort of meals and more, and the space he demanded to pursue his work, without asking much in return.

And so, by November of 1915, with the use of the covariant rule, he brought to a climax his general theory of relativity. During November Einstein delivered four crucial lectures to the Prussian Academy of Sciences presenting to his scientific colleagues the final formulation of his general theory.[13] The third lecture contained two crucial empirical components: it accounted for the anomalous behavior of the perihelion of the planet Mercury, and it deduced the correct prediction of the bending of the light by gravity.

***

---

[9](1876–1936).

[10]Cassidy [25], pp. 100–101; Nathan and Norden [148], Chap. 1, esp. pp. 1–8. The present-day Euro-zone is somewhat the coming to fruition of this 1914 dream.

[11]*Einstein Papers*, Vol. 8, Doc. 39.

[12]Green [81], Introduction.

[13]*Sitzungsberichte, Prussische Akademie der Wissenschaften*; Reports of Proceedings of the Prussian Academy of Sciences, in *Einstein Papers*, Vol. 6, Docs., 21, 22, 24, and 25 (November 4, 11, 18, 25, respectively).

Fig. 14.5 Le Verrier's
discovery of the advance
of the perihelion of
Mercury. The planet's
elliptical orbit around the
Sun slowly rotates along
its axis

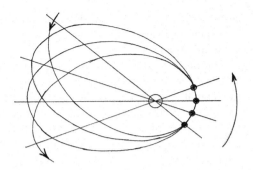

To understand the issue of Mercury, we need to recall Kepler's discovery of the elliptical orbit of the planets mentioned at the start of this Chapter. Kepler's discovery along with other regularities in the motions of the planets, which later were condensed into three planetary laws, were shown by Newton to be derivable from his simple law of gravity. Several generations of the mathematical physicists integrated Newton's physical insights within the developing mathematical system of calculus into a cohesive celestial physics by the early-nineteenth century. It was so sophisticated that when the planet Uranus, discovered by William Herschel in the late-eighteenth century, was found to have anomalous motions in its orbit – speeding up and slowing down with no apparent cause – it was possible to calculate quite precisely where another planet beyond Uranus should be if its gravitational attraction was the cause of the anomaly. In the early 1840s this calculation was independently made by John C. Adams in England and Urbain Le Verrier in France. By 1846, with the empirical confirmation of their predictions, a new planet, eventually named Neptune, was added to the solar system.[14]

Le Verrier went on to become Director of the Paris Observatory where he made a second discovery, important to our story on Einstein. Kepler's planetary elliptical orbit was a closed path fixed in space, but Le Verrier found that the ellipse of Mercury was very slowly rotating such that a looping path is generated as in Fig. 14.5. One way of specifying this motion is by mapping the closest point to the Sun of the planet's orbit (the perihelion) as it moves ahead after each orbit; this very slow motion – only 43 seconds of arc every century – became known as the anomalous advance of the perihelion.[15] This motion was an anomaly in Mercury's orbit, in that it was not explained by Newton's law. Using the same logic he had employed in predicting the planet Neptune, Le Verrier calculated that another very small planet

---

[14]I am avoiding the English-French priority dispute, for which there is a large literature. For one interpretation of the dispute, and the even further claim that the discovery was a "fluke," see Rothman [175], Chap. 4.

[15]To be historically accurate, Le Verrier got 38 seconds, which was later corrected after he died. Also, the actual advance of the planet was available in planetary data probably since the seventeenth century. Le Verrier's calculation involved subtracting the gravitation effects of all the planets to account for this motion, and he was left with the unaccounted, hence anomalous, 38 arc-seconds per century.

must be between the Sun and Mercury to explain the anomaly. Initial results were disappointing for no planet was found at the predicted spot; however, years later, upon hearing that an astronomer had found the predicted planet, Le Verrier was so elated that he gave the name Vulcan (the god of fire) appropriately to the object, hoping to have added another planet to our solar system. Subsequent searches by other astronomers, however, came up empty, since apparently a sunspot was mistakenly seen as the missing planet.

***

This anomaly was still be around without explanation into the next century, which brings us back to Einstein's general theory of relativity in 1915. He found that the slight difference between his theory of gravity and Newton's was just the quantity needed to explain the small advance of Mercury's perihelion. We can hear his elation in a letter to Ehrenfest a few months later: "Imagine my delight at realizing that general covariance was feasible and at finding out that the equations yield Mercury's perihelion motion correctly. I was beside myself for days with joyous excitement."[16] Einstein is also reported to have said that the discovery gave him palpitations of the heart, and when he made the calculation he had the feeling that something snapped inside him.[17]

It is often said that the deduction of the Mercury's perihelion was a surprise for Einstein. But in a letter to Habicht as far back as 1907 he said that he was "working on a relativistic analysis of the law of gravitation by means of which I hope to explain the still unexplained secular changes in the perihelion of Mercury." It clearly was anticipated and not at all a surprise in 1915.[18] Nonetheless, the elation was justified.

The second empirical component of the third lecture is the prediction of the bending of light by gravity. Unlike the calculation of Mercury's orbit, whose confirmation of the prediction was immediately fulfilled, this prediction was yet to be confirmed. As see above, the prediction of gravity bending light was made earlier in the development of theory (in 1911), but the amount of deflection predicted in 1915 was twice that deduced in the earlier paper. Since we now know that the latter is correct, any empirical tests of earlier prediction (which fortunately for Einstein did not transpire in 1914) would have contradicted the theory and perhaps even been seen as falsifying it. How Einstein would have reacted in 1911 to the eclipse experiment falsifying his idea is pure speculation, but surely it would have given him reason for concern, consternation, or worse. As will be seen, this specific prediction was a key element in the history of the theory and especially in his life as a scientist.

Lastly in the fourth lecture he presented his final formation of a tensor equation incorporating the four-dimensional, non-Euclidean explanation of gravity as being

---

[16] Letter of January 17, 1916. Quotation ending "...von freudiger Erregung." Einstein Papers, Vol. 8, Doc. 182.

[17] The references are in Pais [162], p. 253.

[18] Einstein Papers, Vol. 5, Doc. 69.

the warping of space by matter, and, in turn, bringing the equivalence principle to fruition. The journey within a journey had reached a destination. As Einstein said in the quotation I used as an epigraph to start of this Chapter, "…the years of anxious searching in the dark, with their intense longing, their alternations of confidence and exhaustion and the final emergence into the light – only those who have experienced it can understand it."[19] True, true enough.

This destination was, however, was only a temporary termination. As the next two Chapters will show, there were further quests beyond the horizon.

---

[19] I few years ago a controversy arose among historians of science based on the discovery of an exchange of letters between Einstein and the German mathematician, David Hilbert, who began working on the problem after he heard Einstein present it in a lecture. Did Hilbert beat Einstein to the correct answer first? Most historians now say, No – but some are still not convinced. A good summary of the debate, with extensive citations, is in Isaacson [109], pp. 212–222.

Chapter 15
1916: The Great Summation Paper on General Relativity

The November lectures, published in the Proceedings of the Prussian Academy of Sciences,[1] exposed the struggles and breakthroughs of the last days of "searching in the dark." What was needed next was a logically organized and comprehensive summary of the theory for theoretical physicists – and perhaps, posterity – to devour and digest. This Einstein published three months later, in March 1916, producing one of the greatest papers in the history of science.

This landmark paper, "The Foundation of the General Theory of Relativity,"[2] began with an acknowledgement of the work of Minkowski, "who was the first one to recognize the formal equivalence of space coordinates and the time coordinates, and utilized this in the construction of the theory." In addition, Einstein "gratefully" thanked "my friend, the mathematician Grossmann, whose help not only saved me the effort of studying the pertinent mathematical literature, but who also helped me in my search for the field equations of gravitation."[3]

The core of the paper then began with a brief review of special relativity as being restricted to the case of inertial systems, and the subsequent deductions about time, length, and mass. In the second section he asserted that "there is an inherent epistemological defect" in special relativity that was, "perhaps for the first time, clearly pointed out by Ernst Mach." Einstein, clearly, was referring to what he later called Mach's principle, which he explained, concluding, with emphasis: "The laws of physics must be of such a nature that they apply to systems of reference in any kind of motion."[4] This plea to include acceleration within the relativity postulate was immediately followed by the case we deem the elevator thought experiment; his

---

[1] *Sitzungsberichte, Prussische Akademie der Wissenschaften*; Reports of Proceedings of the Prussian Academy of Sciences, in *Einstein Papers*, Vol. 6, Docs., 21, 22, 24, and 25.

[2] *Einstein Papers*, Vol. 6, Doc 30. The document published in the English translation volume is a reprint of the translation in the book Einstein et al. [57], pp. 109–164, except for the first page, which was missing in the book and is reproduced in the *Papers*. Otherwise, I am using the book for my citations of this paper.

[3] *Einstein Papers*, Vol. 6, Doc. 30, p.146 ET.

[4] Einstein et al. [57] [1916], pp. 112–113.

D.R. Topper, *How Einstein Created Relativity out of Physics and Astronomy*, Astrophysics and Space Science Library 394, DOI 10.1007/978-1-4614-4782-5_15, © Springer Science+Business Media New York 2013

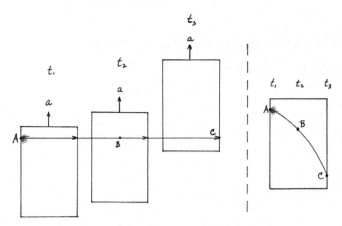

Fig. 15.1 An illustration explaining the bending of light by gravity based on the equivalence of gravity and acceleration. This is a consequence of Einstein's 1907 thought experiment (Fig. 12.1)

specific case is a bit more abstract, but essentially is the same thing, by asserting the identification of a non-inertial system with a gravitational field. "It will be seen from these reflections that in pursuing the general theory of relativity we shall be led to a theory of gravitation since we are able to 'produce' a gravitational field merely by changing the system of co-ordinates."[5] By putting the word "produce" in quotation marks, Einstein was inferring that gravity can be turned on and off (like electricity) as he speculated about in the 1907 thought experiment. In sum, the two paths toward the relativity of acceleration – as traced above in some detail in Chaps. 13 and 14 – are delineated at the start of the long 1916 paper.

In the same paragraph he then deduced the important prediction of the bending of light by gravity as an extension of the elevator thought experiment, for the linear path of a ray of light will appear curved from the viewpoint of an accelerating system. This is clearly seen with the following visualization (Fig. 15.1). The left side depicts one elevator accelerating over three instances of time. (The three instances are spread-out the horizontally, and labeled times t_1, t_2, and t_3 in the drawing, rather having them overlap vertically, so as to make the sequence more clearly seen.) The elevator is accelerating in the direction of the arrows, and the light beam, originating at **A**, is moving straight across the space from points **A** to **B** to **C**, since the beam's motion is independent of the motion elevator. From the viewpoint of a person in the elevator, therefore, the beam moving from **A** to **B** to **C** appears to curve over times t_1, t_2, and t_3 as shown in the composite diagram on the right side of the figure. As a result, if acceleration is perceived as gravity, which the equivalence principle asserts, then the person within the elevator correctly concludes that gravity

---

[5] Einstein et al. [57] [1916], p. 114.

bends light. Hence, near the beginning of this long paper, Einstein put forth one of the key predications of his general theory of relativity.[6]

He went on in a very long middle section to introduce four-dimensional, non-Euclidean, space-time and set-up the equations of motion using the mathematics of tensor calculus. In some ways it was a crash-course in the subject. Putting this in words: he showed that gravity can be explained by an extra-dimensional curvature of space around matter. Specifically he demonstrated that Newton's law of gravity is derived from his tensor equation as a first approximation, which is a very important calculation.[7] He then said – and he later showed – that to a second approximation the explanation of the advance of the perihelion of Mercury follows. "These facts must, in my opinion, be taken as convincing proof of the correctness of the theory," he asserted. Indeed, I would concur with Einstein; it means that his theory encompasses Newton's but also is more comprehensive, which is a very strong reason for gauging the correctness of a theory. It does not mean that Einstein's theory replaces or falsifies Newton's theory, but that the new theory embraces the old, explaining the same phenomena, and more. The "and more" made it a more comprehensive model of reality, and reinforced Einstein's growing ideation of the progress of science as following an evolutionary (not revolutionary) course.[8] In speaking of "a comprehensive model of reality," we are being true to Einstein's notion, for he used the phrase "a closer approximation to reality" himself.[9]

At the end of the paper[10] he deduced four predictions or phenomena. They are, in the order they appeared: the gravitational time dilation, the relativistic or gravitational red shift, the gravitational bending of light, and the deduction of Mercury's perihelion advance. (The first two, recall, were conceived at the end of the 1907 summary paper on relativity.) After obtaining a formula for the gravitational time dilation, he wrote, in the very next sentence: "From this it follows that the spectral lines of light reaching us from the surface of large stars must appear displaced towards the red end of the spectrum."[11] (In 1907 he first predicted this phenomenon as the gravitational redshift of light from the Sun.)

On this prediction he commented in a footnote: "spectroscopical observations on fixed stars of certain types indicate the existence of an effect of this kind, but a crucial test of this consequence has not yet been made."[12] Einstein was ever diligent in keeping abreast of experimentation that supported his ideas, the importance of which was discussed before (Chap. 7).

---

[6] Einstein et al. [57] [1916], pp. 114–115.

[7] Einstein et al. [57] [1916], pp. 142–145.

[8] See, for example, Einstein and Infeld [59] [1938].

[9] Einstein et al. [57] [1916], p. 157. *Einstein Papers*, Vol. 6, Doc. 30: *"Eine der Wirklichkeit näher liegende Approximation erhalten wir…."*

[10] Einstein et al. [57] [1916], pp. 160–164.

[11] Einstein et al. [57] [1916], p. 162.

[12] Einstein et al. [57] [1916], p. 162n.

Next was the calculation of the bending of light by gravity, the curvature being a function of the mass of the gravitational matter and the distance of the light-source from the mass. He calculated the specific case of a light-ray passing the Sun and the planet Jupiter, the former being a deflection of 1.7 arc-seconds.[13]

The paper ended abruptly, with the final calculation being the advance of Mercury's perihelion, which was already known to be 43 seconds of arc per century. Here then was a prediction already confirmed. That was it: end of paper. No summing-up, no review, no profound conclusion, no final sentence. I suspect that in Einstein's mind any weighty commentary should come later, if and when the other predictions are confirmed. Such ruminations are at present premature. There was enough of this sort of thing at the start of the paper.

***

One would think that Einstein would be thoroughly fatigued at the point, the culmination of "searching in the dark," of "intense longing" and "alternations of confidence and exhaustion," until "the final emergence into the light." But no: shortly after completing the general theory, perhaps during the momentous month of November 1915, he began thinking about and planning a popular book to explain relativity to the lay person. In a letter to Besso in January of 1916[14] he mentioned the idea of a book on special and general relativity, although he was having difficulty getting started. Nevertheless, the task was important, if not his intellectual duty: "But if I do not do so, the theory will not be understood, as simple though it basically is." The last phrase I particularly like[15]; it is, in a way, the theme of this book. Relativity is indeed extremely simple, not in the sense of being easy but as being fundamental and minimal. In writing a popular account of his theory he was in good company; both Galileo and Newton wrote popular accounts of their ideas.[16] Einstein completed this manuscript in December of 1916, when he wrote the preface; in the spring it was published as *Relativity: The Special and the General Theory*.[17] The book is still in print, having gone through fifteen editions during his life. I have drawn on this book before (Chaps. 6 and 7) in my discussion of special relativity (it is where he introduced the moving train example) and will do so later in this book.

The effort put into the relativity book still did not sap his energy, for in 1917 he published another landmark paper, this one on cosmology, and which is the starting point and springboard for the next Part (**IV**) in this book. Whether due to this long series of

---

[13] I think that the gravitational redshift may, alternatively, be deduced from the bending of light by gravity using the wave model of light. Since gravity bends light, then gravity attracts light; thus a wave of light being emitted from the center of the Sun would be attracted back towards the Sun as it traveled outward, so when it reached the surface of the Sun the light would be stretched to a longer wavelength, that is, it would be shifted toward the red. I do not know if Einstein ever conceived of this.

[14] *Einstein Papers*, Vol. 8, Doc. 178. Letter of January 3, 1916.

[15] *Einstein Papers*, Vol. 8, Doc. 178, "...so einfach sie im Grunde nun ist."

[16] Galileo [71] [1632]. Newton's [150] [1727] was only published after his death. The details are found in Topper [198], pp. 155–162.

[17] Einstein [49] [1917].

creative efforts or not, Einstein became unwell about this time. Various ailments plagued him: liver complaints, stomach ulcers, and general malaise, which lasted for several years.[18] With his move into Elsa's apartment she slowly nursed him back to health. In February 1919 his divorce from Mileva was finalized, and in June he and Elsa were married.[19] Sandwiched between these two personal events was a scientific one: in May there was a solar eclipse of the Sun that provided the scientific world a means for testing a key prediction of general relativity, that gravity bends light.

***

15.1 Summary

Similar to special relativity, the general theory begins with two postulates:

1. *The principle of the relativity of motion, in this case applied to both inertial and non-inertial systems.*
2. *The principle of equivalence, asserting the identity of gravity and acceleration; described otherwise as inertial mass or inertial force being identical to gravitational mass or gravitational force, respectively.*

As with special relativity, we ask what sort of world exists under these conditions, and we find a world where gravity is due to the bending of space around matter. Such a situation expressed mathematically involves a non-Euclidean geometry of four-dimensional space-time, which can be visualized by making an analogy with the visual experience of two-dimensional beings who experience their space as curved into a third-dimension; hence, our space-time is curved into a fourth-dimension. This fourth dimension contains the time variable and hence it is a space-time curvature for us, who experience a three-dimensional world.

The mathematical expression of this four-dimensional world leads to four predictions of the behavior of matter in our world. They are:

1. *The gravitational time dilation, where clocks (and therefore time itself) run more slowly as a function of the gravitational field-strength.*
2. *The gravitational redshift, where the spectrum of light emitted from stars (such as our Sun) is shifted toward the red.*
3. *Light-rays are bent by gravity, such as a light-beam from a star passing a large mass such as the Sun.*
4. *The advance of the perihelion of Mercury is encompassed by the theory, along with Newton's law of gravity.*

---

[18] Pais [162], p. 525; Fölsing [65], p. 855.

[19] A blunt overview of the triangle among Albert, Elsa, and Mileva, based on recent sources, is in Levenson [132], pp. 141–157.

At the time of publication only the last was empirically confirmed. There was scant but suggestive evidence of a redshift in starlight. The testing of the bending of a light-ray by the Sun was yet to be confirmed. The measurement of the gravitational time dilation was beyond the range of clocks in 1916.

***

The formula for the time dilation in general relativity is

$$(1 - GM/Dc^2),$$

where **G** is the gravitational constant,[20] **c** is the speed of light, **M** is the mass of the gravitational body (planet, moon, star, etc.), and **D** is the distance of the clock along a line from the center of the body. Compare this with the time dilation in special relativity, what I called the quantity (**Q**),

$$\sqrt{(1 - v^2/c^2)}.$$

There is a partial similarity in their forms: we see unity (**1**) minus a fractional term, and there is a **c²** term in the denominator of the fraction, although the special relativity dilation is under the square-root sign. Specifically, in special relativity there is one constant (**c**) and one variable (**v**), whereas in general relativity there are two constants (**G** and **c**), and two variables (**M** and **D**). Looking at specific cases of the general relativity dilation, it is clear that for **M** = 0, the dilation is unity (**1**), meaning that the clock reads the same as a clock in a gravity-free space; that is, there is no dilation, which is analogous to **v** = o in special relativity. As well, as **D** gets very large (say, approaching infinity), the dilation again goes to unity, which it should since the clock is free of the gravitational field of the massive object, and again there is no dilation. The formula also qualitatively implies that as **M** increases and/or as **D** decreases the dilation term increases, meaning that time slow down with increasing mass and/or coming closer to the center. In both cases, increasing **M** or decreasing **D**, the gravitational field strength is being increased, so that gravity slows time in general relativity (just as speed slows time in special relativity). Said otherwise: clocks run slower on Jupiter than on Earth, or a clock at sea level on Earth runs slower than one at the top of a mountain.

The latter implies a possible test of the theory, by comparing two clocks at different altitudes on Earth. Doing the math for the case of two clocks, one at sea level and the other on top of Mt. Everest, we find the gravitational time dilation term is 0.9999995. Notice how close 0.9999995 is to unity (1). Such as small difference in time between clocks was impossible to measure until atomic clocks were invented after the Second World War. Furthermore, recall that in special relativity the quantity of the time dilation for an object moving at 98% the speed of light was 1/5 (1 year to 5 years) or 0.2;

---

[20] We wrote Newton's Law (**I.4**) as a proportion: **F** α (**m** × **M**)/**D²**. This was essentially how Newton expressed it. Not until the late-nineteenth century was it written as an equation, which entailed a constant (**G**): hence, **F** = **G** (**m** × **M**)/**D²**. For some details, see Topper [198], pp. 162–163.

but consider the case of the present Space Station in orbit, which is moving at 17,000 miles per hour. For this we get the same number as the gravitational time dilation above, 0.9999995! In sum, significant relativistic effects in special relativity only appear for speeds near light-speed, so in general relativity the effects arise only for extremely large masses – or with extraordinarily sensitive instruments.

***

At the end of Chap. 12 we briefly explored the role of visual thinking as Einstein's mode of thought, and related it specifically to his two famous thought experiments. Chronologically it did not get us past 1907. What of the epic journey to general relativity in 1915? On the one hand, it could viewed as a more analytical approach because its reliance on abstract mathematics; on the other hand – and I believe more fundamentally – the mathematics is that of geometry, the geometry of the field, and in that sense it can be viewed as a visual expression of the theory. True, four-dimensional space-time is not directly able to be visualized, but it is indirectly seen though two-dimensional analog thinking. In this way Einstein's reliance on geometry – albeit non-Euclidean geometry – correlates more with the visual mode of thinking.[21] This topic will arise once more in the last Chapter, on his quest toward a unified field theory.

---

[21] In some ways this problem is a non-problem in the following sense. Geometry and arithmetic had separate histories. Briefly, and very simply put: Geometry came from the Greeks. Arithmetic was ubiquitous (every culture eventually counts things), but one important branch began in India and coming through Islam evolved into algebra (note the Arabic name). In the seventeenth century, they were put together in what became known as analytical geometry (note the title). A simple example is the equation, $x^2 + y^2 = r^2$, which a circle (geometry) with radius r. Tensor calculus later evolved from these roots.

Chapter 16
1920: Year of Fame, Year of Infamy

Albert Einstein was not "Einstein" until the 1920s. Before entering the world of fame and renown, which included derision and hatred too, he was well-known only within the insular world of theoretical physicists – witness his landing the prestigious position in Berlin. The catalyst for the celebrity status was the eclipse experiment of 1919, organized primarily by the English scientist, Arthur S. Eddington of the Royal Society of London and Director of the Cambridge University Observatory.

The War ended in November 1918 with Germany becoming a republic. The war was over but the excessive nationalism and animosity among European nations remained in many quarters. Eddington, a Quaker and a pacifist, was captivated by relativity and wrote one of the first technical treatises on the subject.[1] Beyond the scientific reason for his effort to confirm the theory, Eddington harbored the ulterior motive of trying to tempering British-German animosity through science. British scientists confirming a German scientist's theory would show that science is an international enterprise and help light the way towards a brighter future between the countries – or so he dreamed. For this he became the driving force within the Royal Society, which launched an expedition to two places where the total solar eclipse in May of 1919 was photographed: in northern Brazil and on the island, Principe, off the West African coast.[2] Although there were numerous problems and difficulties with the expeditions (including poor weather conditions at the time of the eclipse in both places), the scientists came back with plates of images of the Sun and nearby stars. The preliminary results of Eddington's analysis of the data were transmitted to Einstein in September, with the formal announcement made to the press in early

---

[1] Eddington [37] [1923/1924], *The Mathematical Theory of Relativity*. Einstein referred to Eddington's book as the finest on the topic. I remember trying to read through it as an undergraduate, and being first exposed to tensor calculus.

[2] For more on both the pre- and a post-history of using solar eclipses to test Einstein's prediction, see Crelinsten [30]. Our focus is only on the famous 1919 event.

D.R. Topper, *How Einstein Created Relativity out of Physics and Astronomy*, Astrophysics 121
and Space Science Library 394, DOI 10.1007/978-1-4614-4782-5_16,
© Springer Science+Business Media New York 2013

November. Within the limits of experimental error, the data were interpreted as confirming the predicted deflection of 1915, much to Einstein's delight and Eddington's too.[3]

One result surely was the impact it had on Einstein's life. News reporters picked-up the story and with their usual hyperbole turned it into a revolutionary scientific event – which in this case was not far from the truth. They reported that Einstein discovered a new world and placed him beside, if not above, Newton in the pantheon of science. Einstein, Einstein, Einstein.... echoed across the newspapers of the world. Albert Einstein was quickly morphed into "Einstein," where the eponym remains today.[4]

At first he wallowed in the limelight – who wouldn't? Elsa too liked the almost constant attention of reporters and photographers. Once, when asked his profession, Einstein sarcastically said he was a model for photographers. This celebrity status, however, was double-edged; the relentless hounding and lack of privacy over time became trying – all of which was quite annoying. Yet worse were the more nasty variety of attacks on Einstein, some with anti-Semitic overtones. With the rise of anti-Semitism in Berlin, he became friends with a group of Jewish intellectuals who were enamored with Zionism – which reignited his childhood affinity with Judaism,[5] if not also harking back to his friendship circle around Judaic matters during his sojourn in Prague (Chap. 8). Ominously, during 1920 there were a series of episodes that made this previously non-observant Jew even more conscious of his ethnic identity, so much so that we find him increasingly identifying himself as being a member of what he called "the tribe."[6]

There were, to be sure, justifiable questions raised about the theory of relativity, both scientific and philosophical. We saw and will see more of these legitimate

---

[3] For the history of changing interpretations of the veracity of Eddington's experimental results, see Crelinsten [30], Chaps. 5 and 6. It is customary to quote here an often printed story that when asked, what if the experiment had not confirmed his theory, Einstein remarked: "I would have been sorry for the dear Lord. The theory is correct." But Klaus Hentschel [88], convinced me that the story is not true, as I reported in Topper [198]. p. 8. Nonetheless, since writing that I have found this story in Rothman [175], p. 77, who knew the philosopher, Paul Oppenheim, who lived near Einstein in Berlin at the time of the eclipse. Oppenheim told Rothman that when he heard of the positive eclipse result he ran to Einstein to inform him. Einstein's remark was: "It would have been too bad for God had I been wrong."

[4] Much has been written on how he became a scientific celebrity; for example, Friedman and Donley [69]. On this, and how the theory of relativity was seen as being "incomprehensible" not only in the popular press but also among some scientists and engineers, see Crelinsten's interesting two-part article, [28].

[5] Reiser [171], pp. 132–133; Frank [67], pp. 149–158. Recall his short-lived fervent orthodox pre-teen period, mentioned in Chap. 1.

[6] As mentioned before, his first trip to the United States in 1921 was to raise funds for a Hebrew University in Jerusalem, where he saw his role "as high priest and decoy," as he wrote to Solovine. He had been reluctant to go, but as he went on, "I am really doing whatever I can for the brothers of my race who are treated so badly everywhere." Einstein [54] [1921], p. 41. For an overview of Einstein's involvement and attitude toward Zionism, see Ze'ev Rosenkranz [174].

objections. But the hostile and vitriolic tone of the attacks on him in 1920 bespeak of more sinister motives than the scientific search for truth.[7] In the summer an anti-relativity rally, organized by a right-wing political party, was held in the auditorium of the Berlin Philharmonic, where Einstein was accused of plagiarism and propaganda. Certainly the former critique could be leveled against almost any new idea, since no one thinks and works in a vacuum – as surely this book has shown, and in some sense is a theme of the book. The critique of propaganda, however, exposed the more malicious side of this rally, for the organizer of the event identified the sources of the so-called propaganda about relativity as newspapers and publications primarily run by and/or written by Jews. At this rally an etymological dichotomy emerged between "German science" and Einstein's science, which in the 1930s would transform into Aryan vs. Jewish science when the Nazis took power and instated their pseudo-scientific racial theory.

The next episode in the wrangle over relativity took place within the community of scientists at a scientific society meeting in the fall, where an actual debate was arranged between Einstein and Philipp Lenard, the Nobel Prize winning experimental physicist, whose name arose in the love letters between Mileva and Albert (Chap. 3), and whose work Einstein cited in his light quanta paper. Unfortunately we know little of what actually happened in the debate; various individual accounts are mostly contradictory. What we do know is this: it only lasted about fifteen min, and Lenard's primary objections were about the fictitious gravitational fields required by the equivalence principle and the abandonment of the aether. The belief in the existence of the aether was still common among many (perhaps even most) scientists into the 1920s. On the fictitious fields, Lenard raised this question: if the force on a person in a braking train is equivalent to a gravitational field, then where are the masses that are the source of this field? As I interpret this, Lenard's objection cuts directly into Einstein's standpoint regarding Mach's principle. In his reply Einstein referred to distant masses as an answer, which was Mach's explanation. As the next Chapter will show, Einstein began distancing himself from Mach's argument around this time, although I have no evidence that this debate was an important causal factor.

Out of context, this event appears as a mere scientific squaring-off in public, a confrontation that is part and parcel of scientific conferences when a new or radical idea is put forward. But there were other tensions in the background: Einstein was identified as politically a left-wing sympathizer, and Lenard was known as supporting the far right. There also was a social tension in the German scientific community between theorists and experientialists, and even between Berlin and the provincial universities – in both cases, Einstein was indentified with the former and Lenard with the latter.

---

[7] For the following I draw primarily on Van Dongen [203]. See also Frank [67], pp. 158–166 and pp. 250–256; Michelmore [141], Chap. 5; and Cassidy [25], Chap. 6, especially pp. 100–107. For more, and even a reported death threat, see Renn [172], Volume Two, pp. 122–125, and Volume Three, pp. 335–367.

As a possible gauge of the stress resulting from this ordeal, there are a series documented ailments that Elsa presented in the weeks following the debate: she came down with a "bladder infection," "hemorrhages," and "nerves...stressed."[8] Albert and Elsa seriously considered leaving Germany. There was an offer from Zürich, and Ehrenfest in Leiden said he could procure one in the Netherlands. Indeed, Ehrenfest had initiated an appointment for Einstein in 1920 as an adjunct professor at Leiden, whereby Einstein made annual visits lecturing there. On one visit in 1923, during a period when a group of right-wing Germans made death threats against him, Einstein remained with Ehrenfest for six weeks. In the end, however, Albert and Elsa decided not the leave Berlin. Philip Frank reports that when he met Einstein in 1921 at the time of this volatility, Einstein said that within ten years he would no longer be living in Germany.[9]

Needless to say, the growth of what became an anti-relativity movement in the 1920s brought out a spectrum of responses to Einstein's theory. There were legitimate issues raised by serious scientists; there were contrarians (as there will always be; in many ways Einstein himself was one) who looked for an argument without any animosity or hidden agendas; there were those who wanted to prove the theory wrong because there is much prestige in overthrowing something or someone famous; and then there were just plain racists spewing their venom. For example, Lenard's subsequent career under the Nazis exposes the sinister side of his discourse. He joined the Nazi party and continued attacking Einstein as promoting "Jewish physics." Under Hitler he was made Chief of Aryan Physics.[10] Johannes Stark, who recall in 1906 had procured Einstein to write the important summary article on relativity, became aligned with Lenard as a Nazi and an overt anti-Semite. After the Second World War, Lenard lost his post at Heidelberg University, being expelled by the Allied occupation. Stark spent time in prison.[11]

Today there are few objections among scientists to relativity, mainly for reasons to be seen in Chap. 19. There were and there continue to be those trying to prove Einstein wrong. In the past these critiques were put forward mainly in self-published pamphlets and books that often found their way into libraries. Now the Internet has given free reign to any quack with a "proof" that relativity is wrong. And, sad to say, a perusal of anti-Einstein websites, which on the surface appear as scholarly revisionist history, too often reveal upon deeper inspection as being veneer for the same hostile and vitriolic racism that Einstein first encountered in 1920 – that year of fame and infamy.

As I was writing this grim story, I become aware of a new rather bizarre made-in-America sinister attack on Einstein presently being perpetrated by some ultraconservatives.[12] It would hardly be worth mentioning for its downright

---

[8]*Einstein Papers*, Vol. 10, Docs. 154, 165, and 166.

[9]Frank [67], pp. 177–178. It turned out to be twelve years.

[10]A few quotations from his and others writings, revealing their malevolent, racist, and downright despicable ideas, are in Frank [67], pp. 251–256.

[11]How ironic it is that in Einstein's 1905 paper on the quantum theory of light, besides mentioning Planck, the only other references are to the work of Lenard and Stark.

[12]See, for example, Fishbane [64].

stupidity and ignorance, if it were not for the fact that the ideas seem to be taken seriously by a significant section of the public. Believing that Wikipedia is a liberal-biased source, this group has launched their counter-website Conservapedia,[13] which is ideologically motivated, coming from the same fundamentalist religious base that attacks Darwin and evolution. The article on Einstein, beyond the many factual errors, assails him in the same framework as did the anti-Semitic "Aryan scientists" in the 1920s: in essence, they deny any originality of Einstein's ideas, and assert that he borrowed or stole every concept, equation, and theory from others – an absurdity that flies in the face of virtually all scholarship of the last half-century. A mere perusal of the (so-far published) 12 volumes of the *Collected Papers of Albert Einstein* reveals an overwhelming exhaustive display of an active, fertile, and original mind – into, at least, the early 40s of his life. Allow me to quote again Abraham Pais, Einstein best scientific biographer: "Does the man never stop [think-ing]?"[14] The Conservapedia article on Relativity continues the discrediting and trivializing of Einstein's contribution to the theory, such as giving more credit to Poincaré (recall Chap. 8); but, primarily, it concentrates on undermining its verification, arguing that there is a lack of experiential confirmation of most of the theory. At the same time – or, better said, contradicting themselves – they draw on the theory when finding it beneficial to their ideology. Hence they write: "Creation scientists ...have used relativistic time dilation to explain how the earth can be only 6,000 years old even though cosmological data (background radiation, supernovae, etc.) set a much older age for the universe."[15] Enough said?[16]

Returning to the 1920s: along with the infamy came, thank goodness, the fame. As it grew there were honors and metals bestowed on Einstein throughout the decade, one of which was the Nobel Prize in Physics for 1921. As a result, there were trips throughout Europe, as well as to the United States, Japan, China, Ceylon (Sri Lanka, today), Palestine, and South America, where he often gave lectures and received accolades.[17]

---

[13] http://www.conservapedia.com/Main_Page

[14] Pais [162], p. 182.

[15] http://www.conservapedia.com/Relativity

[16] See Will [212]. Serious and certainly not malicious critiques of Einstein, such as Ohanian [155], unfortunately often become ammunition for nefarious attacks on Einstein. On the general issue of the acceptance of relativity, see Brush [18].

[17] Frank [67], pp. 167–201. Recall from Chap. 2 that he spoke on the origins of special relativity and especially the role of Michelson's experiment during his visit to a high school in Chicago and in Kyoto, Japan.

Chapter 17
1922: What is Time? Bergson Versus Einstein … and The Prize

As seen previously (Chap. 9), the time dilation was probably the most difficult concept from special relativity for readers to conceptualize and accept as real. Einstein, we also saw, presented his interpretation of the time dilation in a lecture in Zürich in January 1911. Furthermore, the idea of the twin paradox arose later that year from a lecture by Paul Langevin, and there was considerable discussion on this problem shortly thereafter. The highly regarded French philosopher Henri Bergson[1] was present at Langevin's lecture. This exposed Bergson to the world of relativity and the time dilation.

The paradox was debated further in the early 1920s, after Einstein proposed another solution involving general relativity. He published it a paper in 1918, which, interestingly, was written as a dialogue – a format used by Galileo.[2] Einstein's "Dialogue about Objections to the Theory of Relativity"[3] pits two protagonists: a critic (named Kritikus) and a supporter (called the Relativist) of relativity. They consider the case of two observers with identical synchronized clocks, initially together at rest, who subsequently are separated as one moves away over an interval of time and then returns, bringing them back together. I have illustrated this in Fig. 17.1 with **A** remaining on Earth and **B** making the round trip. Kritikus takes the position that since Einstein's theory implies the equivalence of all inertial systems, then both frames of reference can be deemed to be at rest, and thus time itself must be a relative quantity. An actual time dilation would violate the relativity postulate; thus both clocks should record the same interval of time and be identical. Einstein (in the voice of the Relativist) counters by noting that observer **B**, by moving, experiences a series of accelerations and decelerations in moving away, turning around, and eventually coming back to observer **A**, who, in turn, only experiences a constant

---

[1](1859–1941).

[2]Galileo [71] [1632], who likely was drawing on the ancient Greek philosophers (for example, Plato's dialogues).

[3]*Einstein Papers*, Vol. 7, Doc. 13, pp. 66–75 ET.

D.R. Topper, *How Einstein Created Relativity out of Physics and Astronomy*, Astrophysics   127
and Space Science Library 394, DOI 10.1007/978-1-4614-4782-5_17,
© Springer Science+Business Media New York 2013

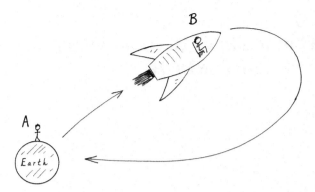

Fig. 17.1 Einstein's resolution of the twin (or clock) paradox. Person **A**, being in an inertial system, experiences a constant gravity. Person **B**, however, is accelerating and decelerating when leaving and returning to Earth, and therefore the clock in this sometimes non-inertial system runs at a slower rate than **A**'s clock

gravitational force while at rest. Therefore their clocks should record different time, with **B**'s clock running slower than **A**'s clock (at rest). Said another way, observer **B**, in experiencing the changes from inertial to non-inertial systems, is really moving, and that is why her clock is actually slower than **A**'s clock. The Relativist (Einstein), in short, is supporting the <u>theory</u> of relativity, not the philosophical position of relativism that Kritikus puts forth.

In his popular book of 1917 Einstein provided a further justification for this argument, if I am reading him correctly,[4] and I am surprised he did not include it in the "Dialogue." After he presented his case for the equivalence between the inertial and gravitational mass, based essentially on the elevator example, he then announced that he "must warn the reader against a misconception" suggested by this thought experiment. Because the person in the accelerating elevator posits a gravitational field that does not really exist, this does not mean that all gravitational fields are only "<u>apparent</u>" ones.[5] It is true that for the case of the person in the accelerating elevator we can choose a different reference frame for which there is no gravitational field. But this possibility is not true for all gravitational fields. He then wrote this decisive case in point: "It is ... impossible to choose a body [frame] of reference such that, as judged from it, the gravitational field of the earth (in its entirety) vanishes." The critical phrase is "in its entirety," for a falling elevator locally on the earth is a case of, so to speak, turning off gravity. This, however, does not apply to the case of the whole Earth, for which we cannot turn off gravity.[6] On Earth we really experience a constant gravitational field, which is the experience of the twin

---

[4]Einstein [49] [1917], p. 69.

[5]Einstein's emphasis.

[6]Although we do explain it by warped (non-Euclidean, four-dimensional) space.

who remains at rest, wherever he is. No contradiction, therefore, arises with the time dilation; and, what's more, no resulting paradox.

As I interpret this argument from Einstein, he is saying that a non-inertial force can always be reduced to a gravitational force, but a gravitational force cannot always be reduced to a non-inertial force. The transformations are not symmetrical. The two forces are not fully equivalents; and if this smacks of absolutism sneaking into the framework, so be it. After all, to say that the twin who leaves the Earth really lives longer implies an absolute difference in age between the twins.

Somewhere I recall reading Einstein as saying that the comprehensive principle of relativity had mere heuristic[7] value for deducing testable results, and once the results were shown to be true, the principle did not need to hold true in all cases.

***

In the spring of 1922 Einstein met Bergson in Paris, the context being a 2-week invitation from Langevin involving a lecture and a series of discussions.[8] Langevin had initially invited Einstein to Paris in 1914, with a lecture planned for the autumn of the year, but the outbreak of war in August led to its cancellation. In the spring of 1922, there was a strong element of anti-German sentiment in France following the First World War and some French nationalist opposed the visit of a German, any German. There was even some resentment from Einstein's German colleagues for making a visit to France, still considered an enemy. At first Einstein rejected the offer, but shortly changed his mind, not wishing to offend his friend, Langevin. He also was encouraged to attend by his friendship with Walther Rathenau, Germany's foreign minister, who saw the trip as part of an effort to amend the animosities generated by the War.[9] Langevin and Charles Nordmann (an astronomer at the Paris Observatory),[10] were concerned about security and thus they escorted Einstein to Paris by meeting his train at the Belgium border. Einstein's hesitation was not baseless; the Society of French Physicists refused to see him; and he cancelled a planned reception at the *Académie des Sciences* upon hearing that thirty scientists planned to exit as soon as he entered the room.

Einstein's did deliver a lecture at the Collège de France on relativity. It was based, in part, on a series of lectures at Princeton University in 1921 and which were published in English as the small book, The Meaning of Relativity.[11] At the time of

---

[7]Remember that Einstein used the term heuristic in the title of this 1905 paper on light for the quantum model, but not in the relativity paper. However, in a different context, he spoke of the 1905 relativity paper as "not to be conceived as a 'complete system,' in fact, not as a system at all, but merely as a heuristic principle...." In this statement, therefore, he does call relativity theory heuristic, but not in the way I recalled above, for here he referring to the rigidity of electrons. The quotation is from the 1907 paper, "Comments on the Note of Mr. Paul Ehrenfest: 'The Translatory Motion of Deformable Electrons and the Area Law,'" in *Einstein Papers*, Vol. 2, Doc. 44.

[8]Frank [67], pp. 194–198.

[9]Grundmann [83], pp. 130–133.

[10]Nordmann published a popular account of relativity in 1922 and later wrote an extensive exposition on the visit. See Nordmann [152].

[11]Einstein [48] [1922].

the Paris lecture, Einstein's old friend Solovine from Olympia Academy days and still a regular correspondent, was working in Paris as an editor, and was, in fact, translating the Princeton lectures into French.[12] The Paris lecture on relativity was followed by discussion sessions (colloquia) the next week at the Collège and the Sorbonne. The latter involved the French Philosophy Society and led to extensive discussions of, not unexpectedly, philosophical aspects of relativity. Solovine was in the audience. So was Bergson.

The concept of time and duration played a key role in Bergson's philosophy throughout his life. Listen to the titles of his early books: Time and Free Will (1889) and Matter and Memory (1907). Thus his 1911 exposure by Langevin to Einstein's radical revision of the notion of time had a profound impact upon him. He rejected facets of Einstein's theory and put forward various objections that culminated in his argument against the reality of the time dilation in a short 1922 book, Durée et Simultanéité: A Propos de Théorie d'Einstein, and in subsequent publications. He asserted that only psychological time was real time, since mathematical time was an illusion. There was no time dilation in the twin paradox for the aging twins, since they aged at the same rate in each individual inertial system. More specifically, the sense of time or duration that each of us experiences in our own reference system is the real, and only real, time.

Doubtless, he conceded, the communication of information between two systems may create the illusion of a time-change. As an analog, consider two people standing apart at a considerable distance in an open space; they perceive each other's size as much smaller than they really are. The distortion, however, is due to the visual diminution in size across space; it is an illusion. Bring them together and they are the same size. The same is true for time: the time dilation is due to the communication of time across space; it is an illusion created by the finite speed of light. Bring the twins back together and they both experience the same duration of time; this means that the durations were the same when they were moving with respect to each other. So the twins are same age when they meet again.[13]

When he met Einstein in Paris Bergson's book was in-press. Bergson initially attended the colloquia just to listen, but a discussion with Einstein inevitably ensued, and ultimately became a debate over their different ideas about time. A transcript of their exchange was published in the Bulletin de la Société Française de Philosophie.[14] Bergson focused on the concept of simultaneity and made the argument that the common sense or absolute concept of simultaneity was not contradicted by relativity theory. Pointing to the fact that one can perceive two or more notes sounded by different instruments in an orchestra as both simultaneous and still hear them separately, he went on the argue that the differences in the time of events as perceived by two observers moving with respect to each other, must still be based on the

---

[12] Einstein [48] [1922], p. 49.
[13] Miller [143], p. 244.
[14] July, 1922; I have used the English translation in Gunter [84].

notion of absolute simultaneity for the observer and the local clock being observed. As a result, "relativity theory contains nothing incompatible with the ideas of common sense."[15]

Einstein countered that individual perceptions are subjective, whereas objective events are independent of individuals, and these events involve a breakdown of simultaneity as deduced in the theory of relativity. Contrary to Bergson, relativity has shown that simultaneity is the illusion, and hence our experience of time is only a subjective mental construction. As he put it, this "psychological time" is "different from the time of the physicist."[16] He repeated this a few years later in a letter to an author of a book on relativity, who also was debating Bergson on this topic. Einstein wrote (the emphasis is his): "It is regrettable that Bergson should be so thoroughly mistaken, and his error is really of a purely physical nature…. Bergson forgets that the simultaneity (like the non-simultaneity) of two events *which affect one and the same being* is something absolute, independent of the system chosen."[17] After reading Bergson's book, Einstein wrote: "Bergson, in his book on the theory of relativity, made some serious blunders; may God forgive him."[18]

On the last day of his visit Einstein, along with Langevin, Nordmann, and Solovine, toured former battlefields and trenches from the War, further revealing the entangled mixture of physics and politics of the trip. That evening, he took the train back to Germany. The Paris sojourn was a trying experience, exposing both political wounds and controversy in physics. He wrote to Solovine that the "days [in Paris] were unforgettable but devilishly tiring; my nerves still remind me of them."[19] In one sense the trip was politically successful, for in early June, there was a German-French friendship rally at the German Reichstag (Parliament) with much applause for Einstein. This hopeful episode, however, was short lived; two weeks later Rathenau's car was riddled with submachine-gun bullets and a hand-grenade thrown in to finish the job.[20] Einstein had serious thoughts that he would be the next Jew assassinated by right-wing thugs. He wrote to Solovine that his life was "nerve-racking since the shameful assassination of Rathenau." He cancelled his lectures and was officially absent from his desk, "though I am actually always here all the time. Anti-Semitism is strong."[21] In October he and Elsa left behind the turmoil of Europe, taking an extended trip to the Far East.

***

---

[15] Bergson, quoted in Gunter [84], pp. 128–133.

[16] Einstein, quoted in Gunter [84], p. 133.

[17] Einstein's letter of July 2, 1924, quoted in Gunter [84], p. 190.

[18] Einstein [54] [1923], p. 59. I should point out that Canales [24] makes a concerted effort at defending Bergson and putting down Einstein's critique. She argues that Bergson did not contradict the twin-paradox and that he was misunderstood by Einstein and other scientists and historians since, which I presume includes me.

[19] Einstein [54] [April 20, 1922], p. 55.

[20] Fölsing [65], pp. 518–519; Isaacson [109], p. 303.

[21] Einstein [54] [July 16, 1922], p. 57.

Bergson's personal dispute with Einstein began long before their actual meeting in 1922, and possibly had repercussions on the granting of Einstein's Nobel Prize. To explain and support this we need first to discuss some aspects surrounding how the Prize was bestowed that are well-know, and some specifics that are much less known.

After the famous eclipse experiment of 1919, Einstein was thrust into the limelight, and relativity becoming table-talk among the intelligentsia; that's why he received his fêted journey to Japan, and as well was a likely candidate for a Nobel Prize. Einstein was awarded the 1921 Prize, but it was not announced until November of 1922 because the committee deferred bestowing the Prize that year. Although Einstein was told that he was granted the Prize before he left for Japan, he did not postpone the trip. So the Prize was announced during his voyage to Japan, and he had to claim it later, in 1923.[22]

The conventional story on the awarding of the Prize is as follows. Einstein received the Noble Prize for his theory of the photoelectric effect, because the Nobel committee thought relativity was still too radical, whereas the photoelectric effect was not. Nonetheless, when Einstein gave his acceptance speech in 1923, he spoke nary on the photoelectric effect but only on relativity, apparently in typical contrarian fashion.

What is less known are several facts accompanying this tale. The original citation for the Prize spoke of Einstein's theory of the photoelectric effect, but it was later changed to just Einstein's equation (that explained the effect).[23] The reason was that the quantum theory of light was still not commonly accepted; indeed, it was seen at the time as being as radical as relativity. It is true that the equation which Einstein deduced in 1905 to explain the effect was confirmed by the American experimentalist Robert Millikan of the University of Chicago in 1915, which was also around the time general relativity was coming to fruition. Einstein's photoelectric effect equation was premised on the light quantum hypothesis. It is not inconsequential that Millikan performed these now-famous experiments in order to disprove Einstein's theory – namely, the quantum of light. Millikan was convinced that the model was wrong, since the interference of light could not be explained without a medium – but to his chagrin the experiment confirmed the equation. In 1917 he wrote about his experimental discovery by speaking of "the apparently complete success of the Einstein equation," even though the physical theory supporting it is "untenable." He explained this with a metaphor: "We are in the position of having built a very perfect structure and then knocked out entirely the underpinning without causing the building to fall. It stands complete and apparently well tested [indeed, by Millikan himself], but without any visible means of support." In sum, an "erroneous theory" has led to an equation that experiments have confirmed – this was Millikan's quandary.[24] In the same year that Einstein retrieved his Noble Prize, Millikan won the Physics Prize for his experimental work, part of which was confirming that equation. In his

---

[22] Isaacson [109], p. 309, debunks the common claim that Einstein only heard of his winning the Prize during his trip to Japan.

[23] Brush [21], p. 218.

[24] Millikan 1917, pp. 229–230.

1923 Nobel speech, Millikan reiterated that "the theory is as yet woefully incomplete and hazy" because "we cannot as yet reconcile [it] at all with well-established wave-phenomena."[25] Again, the interference of light demands a wave, not a particle, model.

Millikan was not alone with this view. The Danish physicist, Niels Bohr, for example, who is often paired with Einstein as the other giant of twentieth century physics, opposed the light quantum model well into the 1920s. Indeed, the word "photon" was not coined until 1926, since there was no need for a name until then.[26] All this, it appears, is reflected in the change of wording of Einstein's Nobel citation from "theory" to "equation." But then, why not give him the Prize for relativity, too? An answer may be found in the presentation speech for Einstein's Prize, which was delivered in 1922, without, clearly, his presence; the German ambassador to Sweden accepted it in Einstein's name.[27] It began with an acknowledgment of Einstein as the most widely known living scientist and then made an immediate mention of relativity. The theory was then put within the context of epistemology, that is, the problem of knowledge – namely, how we come to know the world, and the limitations of this process – and mention was made of a "lively debate in philosophical circles." The next sentence was vital: "It will be no secret that the famous philosopher Bergson in Paris has challenged this theory [of relativity], while other philosophers have acclaimed it wholeheartedly."[28] Mention then was made of the "astrophysical" (really, the cosmological) application of relativity, which Einstein was exploring at the time (and which is the topic of Part **IV**). The presentation went on to mention Brownian motion, and then the remainder and longest part was on the photoelectric effect.

That Bergson, by name, was mentioned at the start, within the context of relativity, for which Einstein was not being given the Prize, is of more than passing interest. It seems that Bergson's critique was taken seriously by the Nobel committee, and was a factor in the choice of the specific justification for the Prize – a choice, seen at the time, between two radical ideas. But there is more.

After returning to Europe from Japan Einstein was slatted to receive his Prize in December 1923, but a committee member suggested instead July, to correspond with the meeting of the Scandinavian Society of Science, to which Einstein agreed. The member also suggested the topic to be relativity theory, to which he also agreed, even though he had intended to speak on his unified field theory.[29] For most of the lecture he acquiesced and spoke on special and general relativity; nonetheless, in the last few paragraphs, he brought-up what he called "the subject of lively interest" – namely "the identity between the gravitation field and the electromagnetic field." He went on to outline his approach to unifying physics, and concluded that "there is reason to hope that a generalization of the gravitational equations will be found

---

[25] Quoted in Brush [21], p. 219n.

[26] Brush [21], p. 223n.

[27] Pais [162], p. 503.

[28] Quoted in Pais [162], p. 510.

[29] Pais [162], p. 504. This significantly modifies the conventional story.

which includes the laws of the electromagnetic field."[30] So, in the end, he, in part, got his way. Or, said another way: he revealed himself to be only a quasi-contrarian – contrary to the convention story of Einstein and the Noble Prize.

One final remark: Pais, who knew Einstein closely, reports that as Einstein got to know Bergson better, he came to like and respect him.[31]

***

This debate between Einstein and Bergson had an afterlife among philosophers for the next several decades, some well-known names in the fray being Alfred North Whitehead, Bertrand Russell, Martin Heidegger, Karl Popper, and Jean-Paul Sartre. Those present-day readers familiar with the so-called Science Wars of the 1990s may recall the name of physicist Alan Sokal, a zealous crusader waging war on the so-called postmodernists who he said believed that science is a mere social construction with little or no truth-value. Sokal traced the "historical origins" of this dispute back through some of the above thinkers and ultimately to the Einstein-Bergson debate.[32]

Throughout my university life, starting in my undergraduate years, I have often been aware of a tension, sometimes surfacing as an animosity, between physicists and philosophers. In the public mind today, the most prominent physicist since Einstein is Stephen Hawking, but among physicists it is likely the late Richard Feynman. A good example of what I am talking about is found in the essay "Relativity and the Philosophers" in his well-known *Feynman Lectures* from the 1960s. A hostile attitude and distain for philosophy can be heard in Feynman's rhetoric: "These philosophers [of science] are always with us, struggling in the periphery to try to tell us something, but they never really understand the subtleties and depths of the problem."[33]

The Science Wars of the 1990s shifted this conflict to physicists versus sociologists and so-called cultural studies theorists.

***

This is an appropriate juncture to make a few remarks about the issue of relativity and relativism, concepts that have come up several times and in various contexts. Let's begin with a statement from the German physicist Arnold Sommerfeld, in his essay for the 1949 tribute to Einstein, in the Schilpp volumes, Einstein: Philosopher-Scientist. Making a forceful case, he wrote that Einstein's "1905 paper has, of course, absolutely nothing whatsoever to do with ethical relativism," and he went on to stress the role of invariance (not relativism) in the theory.[34] From one narrow point of view he was indeed correct. Yet historian of science Loren Graham made this insightful observation.

---

[30] Einstein [38], pp. 489–490.
[31] Pais [162], p. 510.
[32] Quoted in Canales [24].
[33] Feynman 1963, Sect. 16.1.
[34] Sommerfeld [189], p. 99.

Although Einstein maintained that his scientific and his social views were entirely sepa-rate matters, he never answered the question in a definitive way. One can maintain that Einstein's own life was a witness to a relationship between his scientific views and his social ones. His hopes for order, justice, and rational explanation in the social order were too similar to his striving for order, simplicity, and causality in the world of physics to be a mere coincidence.[35]

The engaging book, Victorian Relativity by Christopher Herbert, places Graham's perceptive remark into an even larger context[36] As he convincing shows, there was a pervasiveness of relativism in the culture of the nineteenth century; indeed, con-trary to the common view of the so-called Victorian mind, relativistic ideas were ubiquitous across a range of subjects in intellectual history. In particular, and impor-tantly, Herbert points to parallels between Einstein's scientific relativity and his political views, the latter topic being one about which we find him writing more and more starting in the early 1920s; specifically Einstein drew upon non-absolutist notions (see previous footnote) to support his progressive political ideology. This, along with the evidence from nineteenth century intellectual history, provides a round about way of linking science and ethics in Einstein's world, contrary to Sommerfeld's assertion in the quotation above. In short, Einstein's ethical views were not relativistic in the sense of being nihilistic, but were relativistic from a non-absolutist or non-authoritarian perspective.

If there is a flaw in Herbert's thesis, recall Sommerfeld's second point; namely that Einstein's scientific relativity is essentially non-relativistic and more a search for invariance, which Herbert does not seem to grasp.[37]

---

[35] Graham 1982, p. 131.

[36] Herbert [90]. It is important to realize that the general concept of relativity or relativism as put forth in Herbert's book does not have its source in the denial of truth or entail a shift toward nihil-ism and/or lead to a defense of anarchy; rather it arose in contrast to, or directed against, intolerant forms of absolutism; especially the closed-minded sort of thinking that often pervaded the Victorian age. Significantly, this brings us back to Einstein's relativity, since various and related forms of absolutism would later form the basis of attacks on his theory, specifically from the Nazi move-ment and Soviet ideology starting around the 1920s, and perhaps continuing with the radical anti-Einsteinean neo-conservatives in the USA today. Also relevant to the last sentence is Cassidy [25], Chap. 6, esp. pp. 102–110.

[37] This point came to mind when reading Sheweber's intriguing idea of a "striking parallel" between Einstein's later attempt to unify gravity and electricity and his political involvement with the "one world government" movement. Schweber [180], pp. 96–100. There clearly are several ways of envisioning a unity in Einstein's thought.

Chapter 18
1931: Einstein's First Visit to Caltech

The 1920s, in introducing Einstein to the experience of being a celebrity, culminated at the end of the decade with his second trip to the United States[1] instigated by, of all people, Robert Millikan, who was skeptical of Einstein's particle model of light, even after he himself experimentally confirmed its predicted equation. Furthermore, after the eclipse experiment, Millikan put forward an alternative "plausible" explanation that he hoped would be true: that the bending of light was caused by refraction from solar gases that deflected the light rays.[2] Nonetheless, he respected Einstein, and realized that he was a major physicist of the century.

Millikan had moved from the University of Chicago to California Institute of Technology (Caltech) in Pasadena, becoming its president in 1921, a position he held until 1945.[3] He wanted to attract Einstein to America and especially to Caltech. He was therefore the driving force behind the following arrangement set up with Einstein: annually to spend the winter term (about two months) as a Visiting Professor at Caltech. The change in the weather alone was a positive attraction. In his heart-of-hearts Millikan hoped that eventually Einstein would permanently leave Europe.

The first sojourn was in the winter of 1930–1931. Those accompanying him on this trip were Elsa, his secretary, Helen Dukas,[4] and an assistant, Walther Mayer,[5] who performed many mathematical computations and was nicknamed "Einstein's calculator." Leaving Europe by ship on December 11, 1930 they first arrived in

---

[1] The first was the 1921 fundraising tour for a Hebrew University in Jerusalem.

[2] Quoted in Crelinsten [28], p. 121.

[3] Technically his title was "Chairman."

[4] Helen (Helena) Dukas was hired in 1928 (on a Friday the 13th, which turned-out to be her lucky day) as his secretary to help him with his growing correspondence and other matters of organizing his papers. She remained in this capacity after Einstein's death. As a trustee of his estate (along with Otto Nathan) she essentially controlled the Einstein Archives until it was transferred to Jerusalem. She died in 1982. For a poignant essay on her see Holton [101].

[5] Walther Mayer (1887–1948). Austrian, Ph.D., 1912. He began collaborating with Einstein in 1930. Pais [162], Chap. 29.

D.R. Topper, *How Einstein Created Relativity out of Physics and Astronomy*, Astrophysics and Space Science Library 394, DOI 10.1007/978-1-4614-4782-5_18, © Springer Science+Business Media New York 2013

New York City, where Einstein gave what became a controversial speech on his aversion to the increasing militarism of the time. The contentious point he made was the assertion that governments would have no power to wage war if 2% of the men called up for military service would refuse to serve. Dubbed "the 2% speech," it was a source of much criticism of Einstein's political views. Millikan, who was politically conservative, was embarrassed by the speech and hoped to muzzle Einstein when he got to California, if he could. From New York their ship next stopped briefly in Cuba before proceeding through the Panama Canal and arriving in San Diego harbor on December 30, 1930. Being the luminary he was, Einstein and his entourage were greeted with a marching band, children bringing flowers, and a barrage of mundane questions from reporters. The group was driven to Pasadena the next day, where Einstein was one of the celebrities at the annual Rose Bowl football game parade. He also later met other public figures such as Charlie Chaplin and Upton Sinclair. Of the Chaplin meeting, which took place at the premier of his now-classic film, "City Lights," there survives a memorable quotation, supposedly utter by Chaplin as a crowd was cheering them: "They are cheering me because they understand me; they are cheering you because they do not understand you." Although after the Second World War Chaplin would have serious problems with the US government, such that he would move to England, at the time he was seen as just a harmless actor playing the role of a hapless hobo.[6] Sinclair, however, was a different matter, being a writer identified with radical socialism. Millikan was uncomfortable with their meeting but was not able to thwart it. I suspect Einstein enjoyed the controversy he was stirring up with his left-wing views brought to America. He certainly was enjoying the weather: in a letter he spoke of "loafing in this paradise."[7]

During this first stay in California Einstein meet numerous physicists and other scientists at Caltech as well as astronomers working at the Mt. Wilson Observatory, whose offices are in also in Pasadena, not far from the Caltech campus. There is an often reproduced photo taken right after a lecture Einstein delivered in Pasadena, in which he is posed between six other famed scientists, all standing in front of the blackboard Einstein used during the lecture (see Photo 18.1). The room is identifiable as the library of the Observatory in the Pasadena offices, because there is a large portrait on the wall behind them of George Ellery Hale, founder of the Observatory (see my sketch of Photo 18.1; Fig. 18.1). Einstein had corresponded with Hale in 1913 on a possible method of verifying the gravitational bending of light by Sun. It seems Einstein was trying to avoid the need to wait for the next Solar eclipse, and he thought a measurement could be made, as he wrote, "by day (without solar eclipse)."[8] In his reply,[9] Hale gives several reasons why such a measurement cannot be made, whereas the eclipse method avoids these problems. Further, he notes that

---

[6] Actually there are radical political undertones in many of his early films.

[7] Einstein [56] (February 5, 1931), p. 105.

[8] *Einstein Papers*, Vol. 5, Doc. 477, p. 356 ET, emphasis his.

[9] *Einstein Papers*, Vol. 5, Doc. 483.

Photo 18.1 Einstein at the Mount Wilson Observatory's Hale Library, Santa Barbara Street offices, Pasadena, California, January 1931. From the left: Milton L. Humason, Edwin P. Hubble, Charles E. St, John, Albert A. Michelson, Einstein, William W. Campbell, and Walter S. Adams. Reproduced by permission of The Huntington Library, San Marino, California

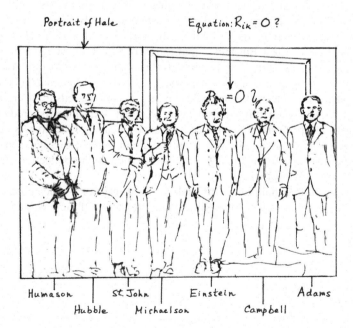

Fig. 18.1 Sketch of the group photograph (Photo 18.1) taken after Einstein's lecture at Caltech in January 1931, with an identification of their names

he has consulted on this matter with the Director of the Lick observatory,[10] William Campbell, who is "interested in the problem," and who was asked to correspond with Einstein. It is true that Campbell was interested in the bending of light problem, as Hale says, but the reason was because Campbell was skeptical of relativity.

Looking at the photo: flanked on Einstein's immediate left is indeed William W. Campbell, next is Walter S. Adams; on Einstein's immediate right is Albert A. Michelson, then Charles E. St. John, Edwin P. Hubble, and Milton L Humason – all of whom played a role in some aspect of relativity.

Michelson had moved to Pasadena to work on experiments measuring the speed of light. As we saw in Chap. 6, the historiographical controversy over the role of the Michelson-Morley experiment on the genesis of relativity theory still lingers among historians. We know that Michelson at the time of the photo harbored serious doubts on relativity since he still was committed to the existence of an aether. Indeed, he was never comfortable with the accolades he received for the aether-drift experiment, which he viewed as essentially a failure; he failed to detect what he believed down deep actually existed. At a formal dinner in Einstein's honor at the Caltech faculty club, Einstein spoke of the Michelson-Morley experiment's role in supporting the theory of relativity. A close reading of the text seems to acknowledge its importance as a later support for the theory but as having negligible influence on its genesis.[11] This runs counter to some remarks made earlier in the 1920s in Chicago and Kyoto, as seen in the last chapter – thus keeping the historiographical issue alive. This trip was the first meeting between them; it was also the last, for Michelson had suffered a stroke about two years earlier and was rather frail, which is visible in the photograph. He died in the spring, not long after Einstein left for Berlin.

Adams was the director of Mt. Wilson, having succeeded Hale. His important experimental work to test Einstein's predication of gravitational redshirt began in 1915 by measuring the burnt-out star Sirius B (the companion of Sirius, the brightest naked-eye star). Sirius B was used because he measured it to be about 50,000 times as dense as water, which meant that it should exhibit gravitational redshift if Einstein was right. By 1925 Adams interpreted his measurements as confirming the prediction.[12]

Campbell had performed a solar eclipse experiment in 1922 to test again the 1919 result; as noted above, he was skeptical about general relativity. In fact, an eclipse experiment performed in 1918 seemed to indicate no deflection. The 1922 result, however, was positive.[13]

St. John, an early collaborator with Hale at the observatory – and another skeptic of relativity – began a series of experiments in 1917 to see if there was a redshift in

---

[10] The Lick is another observatory, further north in the mountains along the California coast, on Mt. Hamilton, near San Jose.

[11] Einstein [40 & 43].

[12] Recent measurements, however, have cast doubts on the validity of Adams work: see Hetherington [92], and Wright, [216].

[13] Crelinsten [29].

the light from the Sun. He first results showed no redshift, as he expected. By 1923, however, he confirmed Einstein's prediction. In the words of historian Klaus Hentschel, St. John was converted to relativity.[14] When Einstein took a trip to the Observatory atop Mt. Wilson, St. John proudly showed him his solar telescope, an encounter that is recorded in several photos from the visit.[15]

Finally, Hubble – whose name is immortalized in the space telescope orbiting our planet as I write this – is standing next to Humason, his faithful assistant; at the time of the photo they were in the midst of a major discovery that would change the course of cosmology. That story, however, is reserved for Part IV. Finally, the meaning of the equation on the blackboard behind Einstein will be explained in Part V.

Knowing what we know so far – such that several of the astronomers in the picture were initially doubters of relativity – this photo of the consummate theoretical physicist among six experimentalists may at once appear as peculiar, humorous, or poignant – depending on the context.

---

[14]Hentschel [89].

[15]There seems to have been only one trip up the mountain to the observatory, as stipulated by Einstein's physician, which according to *The New York Times* was on January 29, 1931.

Chapter 19
Is the Theory True Today?

The essential predictions of general relativity are:

1. Explaining the advance of Mercury's perihelion
2. The bending of light by matter
3. A gravitational time dilation
4. A gravitational redshift
   As well there is the question of experimentally verifying,
5. The equivalence principle

In addition, the experimental confirmations of special relativity (such as the Michelson-Morley experiment) support general relativity, too.

Einstein, as seen, deduced the anomalous behavior of Mercury in 1915. More recent measurements of Mercury's perihelion support general relativity. In the latter years of the last century, such motion was also found among binary star systems, further confirming general relativity.

The bending of light was interpreted a being confirmed by the 1919 eclipse experiment, but historical analysis has cast some doubt on this. Nonetheless, subsequent experiments, especially in the late-twentieth century, using light from quasars passing near the Sun (see next Chapter) have confirmed the theory to extreme accuracy.[1]

The time dilation for gravity has been measured using atomic clocks, for example comparing a clock on Earth with one in an airplane.

The gravitational redshift was the last to be confirmed. The experiment was done in 1960 using gamma rays falling from a 74 ft tower at the Jefferson Laboratory of Harvard University. Even more acute experiments have followed, confirming Einstein's prediction. The previous attempts at measuring the solar redshift are considered inaccurate, since the redshifts are mixed with Doppler shifts (see next Chapter) on the

---

[1]Will [211], pp. 772–773.

D.R. Topper, *How Einstein Created Relativity out of Physics and Astronomy*, Astrophysics and Space Science Library 394, DOI 10.1007/978-1-4614-4782-5_19,
© Springer Science+Business Media New York 2013

turbulent surface of the Sun. As well, the measurements of the dwarf star of Sirius are not precise, without more accurate values of the star's mass and radius.[2]

The experiment interpreted as confirming the equivalence principle is the Eötvös experiment. Variations of this, with accuracy of two the three magnitudes of the original, continue to confirm Einstein's postulate.[3]

My favorite confirmation of relativity involves the construction of the global positioning satellite system (GPS) in the 1970s. It not only provides an application of relativity to today's practical world that anyone can understand, but it involves a rather amusing story.

When NASA began working on positioning 24 satellites carrying atomic clocks in orbit, the mainly military engineers initially thought that using classical Newtonian physics was sufficient. Yet a small group of mainly physicists argued that relativity had to be taken into account, since the satellites were moving at 14,000 km per hour and 20,000 km high. The engineers prevailed maintaining that effects due to relativity would be too small to be a factor. Consequently, the GPS satellites, when first launched, were set to Newtonian physics, but the computers also contained a mechanism to factor in relativistic effects (called a frequency offset) that was turned off. Calculations using both special and general relativity predicted that the speed of the satellites slowed down the clocks by about 7 μs, and the difference in gravity made the clocks in the satellites run faster by 45 μs. The "offset" was 38 μs.

The first measurements for pin-pointing positions on Earth using only the Newtonian calculations were not very accurate. But when the offset switch was turned on, … well, you guessed it. It worked, accurately. The relativistic effects therefore were not trivial, but were necessary for precise GPS positioning.[4] The GPS example shows that relativistic effects are relevant to everyday situations, contrary to the prevailing wisdom mentioned before.

One demonstration of general relativity, which is seldom noted but to my mind is significant, is the experience of astronauts under the condition of free fall. The astronauts in the Space Station orbiting Earth, as they seemingly float in space, confirms Einstein equivalence principle, and hence the general theory too.

Notably it harkens back to Newton's thought experiment, when he conceived for the first time the concept of an artificial satellite. Whether or not the genesis of the idea involved a falling apple – the details are lost in an undocumented part of Newton's life – the idea came about by a mental extension of projectile motion in his mind, seen clearly in the drawing he made to illustrate his thought (Fig. 19.1).[5] Consider a series of horizontal projectiles, say apples, launched from a mountain

---

[2] Will [212], pp. 50–54.

[3] Will [211], pp. 771–772.

[4] Yam [217], pp. 53–55.

[5] I have added to the diagram the case of a body in orbit, which does not appear in Newton's original sketch. The diagram and the experiment, along with some historiographical challenges, are discussed in considerable detail in Topper [198], Chap. 10.

Fig. 19.1 Newton's
drawing illustrating his
thought experiment of a
falling projectile body
going into orbit, if the body
is imparted with enough
initial speed. I have added
the final orbital path, which
is not in his original
drawing

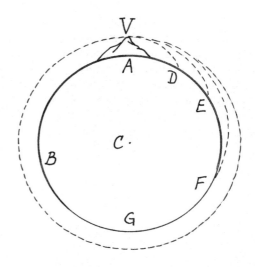

top at the north pole with increasing speed, sending them farther and farther from
the pole; the extreme case is an initial speed sufficient to send an apple all around
the Earth such that it returns to the place of launching, in which case it would go into
orbit. Gravity is still pulling the apple toward the Earth, but the force is balanced
with the opposite centrifugal force, and so it orbits the Earth indefinitely (assuming,
as before no resistance of a medium). It was a monumental intellectual achievement, without a doubt, but was it as far as Newton took the idea. Even though he
tested the equality of inertia and gravity with the pendulum experiment, it was
Einstein who pull them together conceptually within the equivalence principle. Still,
Newton's thought experiment was a key step toward the principle. Since the various
projectiles landing at different distances on the Earth are falling by gravity, then the
special case of the projectile retuning to the launch point and going into orbit is also
falling – falling forever, so to speak. Accordingly, astronauts in orbit are also falling
forever, with gravity directed toward Earth. In fact, their experience is identical to
the person in the falling elevator in Einstein's thought experiment, except in their
case it is really happening. Their experience of floating with the other objects in a
capsule or space station is not exactly a controlled experiment making an accurate
measurement of Einstein's equivalence principle, such as the Eötvös experiment. It
is, however, a qualitative demonstration of the phenomenon, rather like Galileo's
probable demonstration at the leaning tower of Pisa.

In Chap. 10 I pointed to the behavior of the subatomic particles, muons, produced by cosmic rays, as experimental evidence for the time dilation in special relativity. But scientists can also produce muons in the laboratory, particularly in particle
accelerators, moving them at high speeds within circular tracks with electromagnets. Thus at the high-energy laboratory in Geneva, called CERN, muons were

accelerated to 99.7% the speed of light, and their lifespan was extended 20-fold, which was within a 2% accuracy of Einstein's prediction.[6] Since these muons were moving around the track, coming back to their origin and obeying the time dilation, this could be interpreted resolving the twin paradox. "High-speed travel <u>does</u> keep you young," as one as popular writer put it.[7] I am inclined to agree that the CERN experiment resolves the twin paradox.

[6]Will [212], p. 271.
[7]See Calder [23], p. 158, who introduced me to the idea of interpreting the CERN experiment in this way.

Part IV
Cosmology

I have perpetuated something again ... in gravitation theory,
which exposes me a bit to the danger of being committed to a
madhouse

(Einstein to Ehrenfest, February, 1917).[1]

Einstein was a physicist, a theoretical physicist. He was not an astronomer. We know he had a portable telescope during the Berlin years,[2] but as far as I know he never systemically studied the sky. There are also photos of him seemingly peering through telescopes at various observatories throughout the world during his many travels, but most, if not all, are posed publicity shots. Unless I find evidence to the contrary, I believe that Einstein never lost a night's sleep at a telescope.

His undergraduate physics and mathematics courses at the ETH included semester courses in introductory astronomy, astrophysics, and celestial physics, these in addition to the core physics curriculum: mechanics, optics, electricity and magnetism, and so forth.[3] Otherwise he had little contact with matters related to astronomy. From 1909 to 1914 his teaching duties consisted of courses in mechanics, the kinetic theory of gases, electricity and magnetism, and thermodynamics. In Berlin, he lectured on relativity, quantum physics, and statistical mechanics.[4] To repeat, Einstein was a physicist, not an astronomer. In spite of this, in 1917 he initiated a revolution in cosmology, setting in place the theoretical foundation of the subject that endures in present-day cosmology.

---

[1]*Einstein Papers*, Vol. 8, Doc. 294; Pais [162], p. 285.

[2]Bucky, 1992, pp. 51–52, reports a large telescope next to Einstein's desk in his study in Berlin, from which he "observed the night skies around Berlin." The report of the telescope is true, as seen in a photograph of the room in Renn (ed.), 2005, Volume One, p. 261, and Volume Two, p. 114 (identical photos, but with different dates: 1929 & 1927, respectively) where a refractor of about four to five feet, mounted on a tripod, is propped-up by beside his desk and pointing to the ceiling.

[3]*Einstein Papers*, Vol. 1, Appendix E. pp. 362–369.

[4]*Einstein Papers*, Vol. 3, ET, Appendix B, pp. 598–600.

Chapter 20
Cosmological Conundrums and Discoveries Since Newton

Cosmology is key branch of astronomy, dealing with questions around the structure of the universe. The ancient cosmos – systematically codified by Aristotle, and later given empirical support, especially by Ptolemy – was geocentric, geostatic, and finite. Based on a common sense view of the world being as it appears to our senses, the ancient model prevailed well into the seventeenth century. The subsequent scientific revolution, however, bequeathed to the eighteenth century, and after, a radically different cosmic model. The radical change came in two stages. First Copernicus in the fifteenth century moved the Sun to Earth's previous place at the center of the universe, an idea adopted by Galileo, Kepler, and a few other key thinkers up to Newton. The second stage, often called the "breaking of the sphere," replaced the sphere of a few thousand stars at the edge of the finite universe with myriad stars extending into an infinite universe, filled with Newton's invisible gravity, and with our Earth being the third planet from the Sun in our solar system somewhere within that Euclidean space. Two planets were added to our solar system (one in the eighteenth and one in the nineteenth centuries), but the overall structure remained essentially as conceived by Newton when he died in 1727. This was the universe Einstein was born into in 1879.

Then, in the twentieth century, cosmology underwent another revolution, as radical as that of seventeenth century, and its theoretical origin was a 1917 paper by Einstein, whose education and career had little or no contact with a deep study of astronomy. To grasp the significance of his theory, however, we need to look closer into the cosmological and astronomical context from Newton into the 1920s, not long after Einstein's paper was published. As with the previous Parts in this book, we need to start with some background history, which the reader with the heebie-jeebies may skip.

***

D.R. Topper, *How Einstein Created Relativity out of Physics and Astronomy*, Astrophysics and Space Science Library 394, DOI 10.1007/978-1-4614-4782-5_20,

Newton's universe asserted both a universal law of gravitation and an infinite (or, at least, an indefinite) free space filled with the force of gravity.[1] This attractive gravitational force acted among all the stars in the universe, and therefore within any finite space the stars could eventually merge together into one mass. To prevent such a gravitational collapse, the Creator, according to Newton, spread-out the stars into the vast space far enough apart so that the mutual attraction was too weak to pull them all together into one mass. (Since gravitational force decreases as an inverse-square law, the attractive power is miniscule at very great distances.) In the last (third) edition of his *Principia*, written one year before he died, Newton added this sentence near the end of the book: "And so that the [universal] system of the fixed stars will not fall upon one another as a result of this gravity, he [God] has placed them at immense distances from one another." God made the universe larger enough to prevent a gravitational collapse.

Newton's physics and the corresponding infinite model were adopted in the waxing years of the eighteenth century. This cosmos – filled throughout with the power of gravity – was otherwise materially empty, and thus could be conceptualized as a three-dimensional Euclidean space, amenable to the rules of geometry, and therefore to the mathematical physics of Newton.

The most important astronomer of the eighteenth century was William Herschel,[2] a German musician transplanted to England (following in the footsteps of George Friedrich Handel) but who, instead of making a career in composing, fell in love with astronomy – so much so that he eventually made the best telescopes in the world. With these superb instruments he and his faithful sister, Caroline,[3] peered into the far reaches of this newly conceived deep space. In 1781 he became the first person to discover a planet otherwise invisible to the human eye, which eventually became known as Uranus. This discovery not only secured him a place in the history of astronomy, but also secured a nice salary (with a bonus for Caroline) from the king, since Herschel originally named the planet in the king's honor (George). More important for the story of cosmology, however, is another category of celestial objects, quite distinct from planets.

Ever since humans began systematically to watch the night sky, a few strange star-like objects (in both hemispheres) were noticed. Appearing as non-twinkling or blurry stars, the ancient Greeks called them nebulae (i.e., blurs or nebulous objects). One of more conspicuous nebula in the Northern Hemisphere is in the constellation Andromeda; known as the Andromeda nebula, we shall see it play a key role in our story. Since there are so few of them, nebulae were of scant interest to astronomers, although there was the nagging question of where they were placed in the heavens. Being fixed among the stars, they seemed to be within the stellar sphere; however,

---

[1] The distinction between infinite an indefinite was essentially a theological one. Only God was infinite, and if the cosmos were infinite too then God would not be transcendent; this was essentially pantheism, a form of atheism at the time. Calling the universe indefinite got around this problem. Koyré [120], *passim*.

[2] (1738–1822).

[3] (1750–1848).

being misty-looking, they appeared as small cloud-like entities, and therefore should be below the Moon, as were all cloudy and changing objects (such as the Milky Way) according to the ancient astronomers. After Galileo discovered with his telescope that the Milky Way was not a sub-lunar cloud but was really a massive collection of stars ("... a congeries of innumerable stars distributed in clusters"), he turned his telescope to the nebula in Orion's head and discovered ("... what is even more remarkable ...") that it too was collection of stars, much smaller than the Milky Way yet still a cluster of stars that appeared blurred to the unaided human eye. He made the inference (really a conceptual leap) that all nebulae were stars clusters.[4] He proposed, therefore, that nebulae resided among the stars, not below the Moon, and this set in motion the search for further nebulae and the confirmation that they too were star clusters. (Contrary to Galileo's leap, all nebulae are not star clusters; some are composed of a luminous fluid which later was found to be gaseous.)[5]

After Galileo, and as more nebulae were discovered and some could not be resolved by telescopes as being clusters of stars, the issue of their composition arose. From Galileo to when Herschel crossed the channel, over a hundred nebulae were discovered. William and Caroline then catalogued about 2,500 more. On the nature of these objects – are they stars or a luminous fluid? – William vacillated. Early on he thought they were fluid-like, but he later switched to Galileo's point of view after viewing some nebulae as star clusters; he inferred that the blurry ones would eventually be resolved into star clusters, as bigger and better telescopes are made. For him, as for Galileo, the rule was that one hypothesis must fit all. I believe this was due, in part, to the conceptual framework of Newton, who postulated a homogeneous universe. In November of 1790, however, Hershel viewed a "cloudy star" such that "the nebulosity about the star is not of a starry nature" (the emphasis is his); he called it a "planetary nebula" (a term still used today for what we now know is an old star shedding its surrounding shell of gases). Realizing that luminous fluids could therefore exist in space, he cautioned: "Perhaps it has been too hastily surmised that all milky nebulosity, of which there is so much in the heavens, is owing to starlight only."[6] Accordingly, the nebulae that could not be resolved into star clusters Herschel decided were just milky fluids.

Accompanying these questions surrounding the nature of nebulae was an important deduction about the Milky Way. By carefully studying the changing distributions of the stars in the Milky Way over the course of a year (think of our different viewpoints of the Milky Way at the four seasons), Herschel was able to combine conceptually these images from a heliocentric framework into one image, and he concluded that we (within our solar system) are inside the Milky Way. It was not sometime that was out there; instead it was a massive collection of stars which we viewed from within, seemingly at the center. It was also relatively flat, and that was why when we

---

[4]Galileo [72] (1610), p. 62.

[5]There are historical controversies about Galileo's motivations and his perceptions of the nebulae. I discuss some of this, with references, in Topper [198], pp. 62–64.

[6]Quoted in Bartusiak [8], p. 45.

Fig. 20.1 Stellar parallax. If
the Earth moves around the
Sun, then there should be a
semi-annual shift in the
stars, as illustrated by the
geometrical arrangement

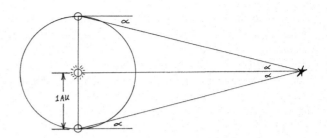

look north in the northern hemisphere (and south in the southern) we see fewer stars, since we are directing our gaze away from the massive star collections in-between. Herschel's reintroduction of a kind of heliocentrism, with us at the Milky Way's center, was possibly cosmic comfort to some, but in the larger scheme of things, where were we? That is, what was the Milky Way's place within the universe?

Herschel vacillated on this too, changing his view several times. The nebulae provided the fulcrum around which his position see-sawed. Were they, for example, external to the Milky Way, such that our Milky Way would look like a nebula from cosmic distances? Since our Milky Way was often referred to by the original Greek term, <u>galaxia</u> (milky-like), these external nebulae were sometimes called galaxies, and thus another image of the cosmos was formed: an infinite space, with galaxies (composed of copious stars) distributed throughout, with our Milky Way being but one. The alternative model: if nebulae were not composed of stars, but actually luminous fluids within the Milky Way, then our galaxy (the sole galaxy) was the entire universe of matter, with perhaps only empty space beyond. At the death of William Herschel in 1822 there was no consensus among astronomers whether nebulae were or were not external to our Milky Way.[7]

An important discovery in 1838 impacted on this question – the first measurement of stellar parallax. In Fig. 20.1, if the Earth moves around the Sun, there should be a semi-annual (parallactic) shift in a star of 2α degrees. The 1838 measurement of 2α for the star 61 Cygni was found to be 0.3 arc-seconds.[8] This meant not only that the Earth actually moves (a confirmation taken for granted by this time) but more importantly it provided a way of directly measuring stellar distances using a triangulation technique. The distance of 61 Cygni in what are called astronomical units (**AU**, setting the Earth-Sun distance to 1 unit) was 700,000 **AU**. As more and more stars were found to have parallactic shifts, the measured stellar distances got

---

[7]Hoskin [103], pp. 231–255.

[8]This designation refers to star number 61 in the constellation Cygnus the Swan. To picture the scale of this specific parallax, consider this Earthly analog. As I write this I am sitting in Winnipeg in the middle of Canada, about 100 km from the Minnesota/North Dakota border. Assuming the Earth to be flat, and my having the ability to see due south as far as Mexico City, a mini-van sitting there perpendicular to my line of sight would encompass this angle of arc!

larger and larger, such that a new unit was required. The new calibration was based on using the distance light travels in year, and thus the unit "one light-year" was introduced, with 61 Cygni's distance being eleven light-years.[9]

This parallactic technique was a powerful tool, for previously only distances within the solar system alone were directly known. Now calculations could be made beyond the last planet (Neptune at the time).[10] Yet the technique had a limit, primarily due to the size of the Earth's orbit as the base of the triangulation. The limit was somewhere under 500 light-years. Even larger instruments – the largest telescopes in the world by the mid-nineteenth century were built in England and Ireland – could not break beyond this geometrical limit.

These massive telescopes did, however, reignite the issue of the nebulae, for they revealed for the first time a spiral shape of some nebulae, which were often conceived of as proto-solar systems; that is, as swirling gases[11] evolving into other solar systems, and this revised the possible dual nature (gases and stars) of nebulae. In addition, the number of nebulae catalogued grew in the nineteenth century to over 100,000.

By the end of the century it was generally conceded that the empirical limit of the measurable universe for us at the center of our Milky Way was a sphere of radius about 500 light-years. Of utmost significance to the cosmic question was this realization: the 500 light-year limit was within our galaxy. For this reason most astronomers in the early twentieth century affirmed that we were empirically trapped within the Milky Way. Everything we saw was confined to our galaxy. Philosophers and visionaries could speculate about other galaxies deep in space, but such notions were, at best, science fiction. Astronomers (most of them, at least) knew better. Only a few did daringly maintain that some nebulae could be other galaxies, an idea that became known as the island universe hypothesis – as if our Milky Way were merely one of myriad such galaxies, spread-out like islands throughout the vast universe.

Certainly the discovery of stellar parallax was of major significance in our capacity to measure the universe, to find our place in the scheme of things. Yet of equal magnitude was this question: what are these things of which we speak, or in short, what is the universe made of? An instrument that was able to furnish answers to this question was invented in the latter half of the nineteenth century – the spectroscope. Ever since Newton around 1666 directed a beam of an image of the Sun through a window in a darkened room, sending the beam through a prism and onto a blank screen to produce the spectrum of light, this rainbow of colors was the object of study and analysis. As seen in Chap. 3, in the early part of the nineteenth century the spectrum was discovered to extend into humanly invisible realisms, first beyond the red into the longer wavelengths, and at the opposite end into shorter ones.

---

[9] A light-year is about 63,000 **AU**. Therefore 700,000/63,000 = about 11 light-years.

[10] Pluto was not discovered until 1931.

[11] The actual gaseous constitution of the nebulae, as opposed to just a luminous fluid, was discovered with the invention of the spectroscope, mentioned shortly below.

The spectroscope, described simply, consisted of a prism and small telescope, such that Newton's experiment was repeated with further accuracy.

Used first as a terrestrial instrument, the spectroscope revealed that the spectrum of colors of heated elements (say, hot gases, burning salts, or "red hot" metals) produced color patterns unique to each chemical element. This meant that the chemical compositions of objects could be found by merely looking at their spectrum. The application of this to astronomy contradicted what was sometime seen as mere common sense: except for our intimate knowledge of the substances composing our Earthly world, we can never know the stuff beyond, unless we can travel out and bring it back; seeing alone cannot reveal what the stars are made of. Not so: a spectroscope could detect the chemical composition of a distant object by just drawing-in its light spectrum. Indeed, some of the first spectra of nebulae indicated that they were composed of gases alone, which not only explained what the cloudy luminosity was, but implied that they were probably within our galaxy and rebuffed further the island universe idea. At most, nebulae could be proto-solar systems. By the end of the century observational astronomers had two powerful and essential instruments, spectroscopes and telescopes, able to probe the what and the where of the universe, respectively.

Beyond these matters of theory and experiment, astronomy in the twentieth century saw the start of an important geographical shift in the center of science from Europe to the United States, initially in observational astronomy. This was fueled primarily by vast sums of money from wealthy private donors – those superrich American capitalists turned philanthropists – who proudly built the best observatories and the biggest telescopes in the world. Many were placed atop mountains along the California coast (Mt. Hamilton, Mt. Wilson, Mt. Palomar) – for the clear and fresh air, as it then was – as well as in Arizona (the Lowell observatory), Wisconsin (Yerkes), and New England (Harvard). Key discoveries in observational astronomy during the first two decades set the stage for the cosmic revolution that followed in the 1920s.

The first important discovery was based on the work of Henrietta Leavitt at Harvard Observatory between 1908 and 1912.[12] Poring over photographic plates of stars, she discovered a class of stars, initially in the constellation Cephius (the king), that had regular periods of fluctuation in brightness (or luminosity), which became known as Cepheid variables. Their periods of variation in brightness ranged from a few days to several months, but always with a fixed number for each given star. Finding first sixteen Cepheid variables and later nine more within the small Magellanic cloud,[13] she discovered a correlation between their average brightness

---

[12](1868–1921). She graduated from Radcliffe College with a degree in science (and astronomy) and worked at Harvard Observatory, which hired women to catalogue stars on photographic plates.

[13]Visible in the southern hemisphere and named after the Portuguese explorer Magellan, the two Magellanic Clouds are the most conspicuous nebulae anywhere visible to the human eye.

Fig. 20.2 Leavitt's law.
The period-luminosity
correlation she discovered
and published between
1908 and 1912

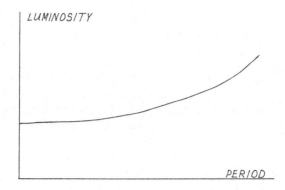

and their period: the brighter stars had longer periods, and vice versa.[14] Since all the stars in the cloud were the same distance from the viewer, this meant that their relative brightness was intrinsic to each star, not a function of their different distances from us. The graphic representation of this correlation is seen in Fig. 20.2, where the vertical axis is the relative (average) brightness of the stars (called luminosity) and the horizontal axis is the period. Known as the period-luminosity law, and found today in any astronomy textbook, I like to call it Leavitt's law.[15]

The significance of this was realized by the astronomer Harlow Shapley, working at the Mt. Wilson Observatory.[16] Although no Cepheid variable was close enough to measure its distance by parallax, Shapley attempted to calibrate the vertical axis so that it denoted absolute values by another, less direct and less exact, method – namely, using the relative lateral motions of eleven Cepheid variables. (Think of watching the landscape passing by from a moving vehicle: the nearby fence posts move past more quickly than a farmhouse at a distance, which in turn moves faster than the distinct mountains, and so forth. All things being equal, the same correlation should apply to moving stars, and thus their real distances may be calculated.) In this way Shapley turned the vertical axis into absolute values. As such, this period-luminosity law was the most powerful tool yet devised for measuring stellar distances, for it broke the 500 light-year barrier. Here is how it works: measure the period of a Cepheid and, using this graph, its intrinsic (absolute) brightness is derived; then compare this actual luminosity with the

---

[14]Leavitt [127, 128]. She noted the "unusual difficultly," in resolving precise data for these stars due to their crowded distribution, their faintness, the shortness of their periods, and other factors. After first measuring sixteen variables, she commented that it "is worthy of notice that ...the brighter variables have the longer periods" (1909), p. 107. Adding nine more, she said about the twenty-five variables: "A remarkable relation between the brightness of these variables and the length of their period will be noticed" (1912), p. 1. Remarkable, indeed.

[15]We now know that there are many types of variable stars, but Cepheid variables are useful because they are easily recognized from the way their luminosity varies with time – a quick rise is followed by a slow decline.

[16](1885–1972). Shapley was one of the key American astronomers of the early twentieth century.

perceived brightness; from this the star's distance is calculated because the intensity of light decreases by an inverse-square law.[17]

Shapley then used Leavitt's law to measure stellar distances within the Milky Way and deduced two things: its size and our place in it. For the latter he found, contrary to Herschel, that our Solar System is near the edge of our galaxy, which is true. His measurement of its size, however, was 300,000 light-years in diameter. (This was off threefold, due to his inexact method of calibrating absolute luminosity; in addition, interstellar dust made objects appear dimmer, and so he thought they were farther away than they really were. Not until 1930 was the measurement corrected to 100,000 light-years by astronomers working at the Lick observatory on Mt. Hamilton.)[18] Knowing – or least potentially knowing – our galactic size, and now with a tool (Leavitt's law) making it possible to measure beyond the limit of parallax, some astronomers, not surprisingly, speculated again about the island universe hypothesis. Since we were no longer empirically trapped within our galaxy, some minds wandered (and wondered) beyond into deep space.

Shapley, not unexpectedly, thought the 300,000 light-year size meant that everything we see still resides within our very large galaxy, and this included all nebulae. Now those nebulae that are really star clusters are obviously within the Milky Way, but so too, he thought, are the gaseous ones, spirals and others. As he wrote in January 1918, "there is no plurality of [island] universes .... The (Milky Way) galaxy is fundamental in what we call the universe."[19] Indeed, the tale has been told that an assistant came to him one day with a photograph of the Andromeda nebulae on which some markings were made, which the assistant thought might be Cepheid variables, and Shapley immediately rubbed them out saying that no stars could exist in this nebula since it was made of gases alone. If this undocumented story is true,[20] it came to haunt Shapley later, as shall be seen.

About the time that Leavitt was completing her work on Cepheid variables another celestial puzzle was unfolding. In late 1912, at the Lowell observatory in Arizona, the astronomer Vesto M. Slipher[21] made a parallel discovery. Believing that nebulae were proto-star systems, he was measuring their light through a spectroscope.[22] Starting with the nebula in Andromeda, he found its light was shifted toward the shorter blue-violet wavelength (what became known later as blueshift). Doing the same for other nebulae he found most of their light shifted instead toward

[17]Incidentally, this was discovered by the astronomer Kepler in the seventeenth century, who also published important work on optics.
[18]Bartusiak [8], pp. 128–129, and p.133.
[19]Quoted in Bartusiak [8], p. 129.
[20]Topper [198]. pp. 195–196.
[21](1875–1969). Slipher was an American astronomer, who worked at the Lowell Observatory his entire career.
[22]Perceval Lowell, founder and director of the observatory, believed nebulae were proto-solar systems, and he set Slipher the task of measuring their spectra to confirm this hypothesis.

the longer red wavelength. In August of 1914 he presented his finding to the American Astronomical Society (AAS) meeting, announcing that twelve nebulae exhibited redshift and three blueshift. By 1917, twenty-one were shifting toward the red and four toward the blue. What did this mean?

Today the immediate answer is that they were Doppler shifts, caused by their motion either away from or towards us. In mid-nineteenth century, the Austrian physicist Christian Doppler[23] discovered that sound waves in air change their wavelength (and therefore their corresponding pitch or note) depending on whether the source of sound moves toward the listener (thus squeezing the wave and raising the note), or moves away (thus stretching the wave and lowering the note). This was empirically demonstrated using a railroad flatcar carrying several trumpeters hitting the same note while riding past a station.

One question that arose from Doppler's discovery was whether there was a corresponding effect for light, since light too was a wave. There were two reasons at the time why there should not be a Doppler shift for light. First, light is a wave in aether, and sound a wave in air, these being different media. Second, sound waves are longitudinal waves (such as a vibrating spring) whereas light waves are transverse (such a vibrating string). These essential differences delayed the immediate deduction of applying to light what was called at the time the Doppler principle. In 1905, despite the postulation of a particle theory of light in his photoelectric effect paper, Einstein, in the relativity paper, deduced an optical Doppler effect from his electromagnetic revision of Maxwell's equations. As special relativity was slowly adopted by physicists, the optical Doppler principle came along too. Nonetheless, at the time of Slipher's work there was no consensus on the question.

Slipher reluctantly (or perhaps hesitatingly) assumed a Doppler shift for the nebulae he found. This meant that the redshifting nebulae were moving away and the blueshifting ones towards us. Noting further their spatial distribution, he found that most of the red nebulae were on one side of our galaxy; Andromeda and most of the other blue ones were on the opposite side. This he took to mean that our galaxy was drifting in one direction with respect to these nebulae, which further implied that these nebulae were (or eventually would be) external to our Milky Way. Such a framework gave some credence the island universe idea.[24] Despite this work, in January of 1918, Shapley declared, as seen above, that "there is no plurality of universes" based on his measurement of the extreme size of our galaxy. By the 1920s, astronomers' opinion on our place in the universe was confused and contradictory.[25]

When Slipher first publically presented his discovery of the red- and blue-shifting nebulae at the August 1914 AAS meeting, in the audience was a research assistant from the Yerkes Observatory of the University of Chicago – Edwin Hubble.[26]

---

[23](1803–1853).

[24]Dewhirst and Hoskin [33], p. 327.

[25]Bartusiak [8], pp. 77–89; Topper [198], pp. 193–4.

[26](1889–1953). Of course, the month of August 1914 was auspicious in another way: the start of the First World War.

After completing his Ph.D. in 1917 Hubble joined the war effort; on returning from Europe in 1919, he landed a job at the Mt. Wilson Observatory. Shapley was still there, but moved in 1921 to become Director of the Harvard Observatory.[27] In those few years together at Mt. Wilson there was little interaction and collaboration between Hubble and Shapley, who had very different personalities.

In October 1923 Hubble found a Cepheid variable in the Andromeda nebula, and using Leavitt's law he measured its distance. The result he obtained was a distance of one-million light-years. (The actual distance is near two-and-a-half million light-years. The error was due to his using Shapley's erroneous calibration; in addition, the Cepheids Hubble used were more luminous than those used by Shapley in measuring the Milky Way, and they therefore were farther away than he thought, so his calculate distance to the Andromeda nebula was too small.)[28] All this was not known at the time, but nonetheless, qualitatively the result was significant, then and now. More than significant: if true, it was astonishing – it meant that indeed the Andromeda nebula was external to our galaxy. Hubble hesitated announcing this result to the astronomical community without further evidence; he wanted to find other Cepheid variables in the nebula, and possibly other nebulae external to our galaxy. He did, however, immediately inform Shapley in a letter. A student who was present when Shapley first read Hubble's letter reports that he exclaimed: "Here is the letter that has destroyed my universe."[29] Perhaps Shapley also was recalling those markings of possible Cepheid variables he had rubbed-out on the Andromeda photograph years ago.

Hubble was a committed empiricist, reluctant to speculate beyond raw data – thus insisting on extensive evidence before formally presenting his findings to the scientific community. Still, as rumors of Hubble's confirmation of external nebulae spread among astronomers, he was pressured to make public his findings, which he eventfully did by sending a summary of a paper he was writing to the AAS meeting of January 1, 1925. Hubble's paper, read by someone else, revealed twelve Cepheid variables in Andromeda and twenty-four in a nearby nebula, thereby indicating that they both were about one-million light-years away. The paper won a prize and put the island universe idea front and center in the astronomical world of the mid-1920s, as our Milky Way appeared to be just one of many more galaxies within a universe of perhaps innumerable galaxies, as more and more nebulae turned out to be galaxies too.

---

[27] Shapley arrived at Harvard on April 1, 1921. His appointment was a relief for Leavitt, who was stifled from performing further research on Cepheid variables by the previous Director who dictated what she could and could not do. Encouraged by Shapley, she was now free to go back to her beloved Cepheid's. But sadly it never went very far; she died of stomach cancer on December 12, 1921 (Bartusiak [8], p. 99). Having been present at her death, Shapley speaks of her in his autobiography as "one of the most important women ever to touch astronomy" (quoted in Topper [198], p. 195).

[28] Bartusiak [8], pp. 203 and 259.

[29] Payne-Gaposchkin [165], p. 209.

Chapter 21
Einstein 1917: Modern Cosmology Is Born

Meanwhile, a parallel story was unfolding in the rarified world of general relativity, which at the time was a branch of theoretical physics mainly isolated and mostly unknown to empirically driven astronomers working with their large telescopes, spectroscopes, and other real-world gadgets.[1]

Einstein, the theorist, published his paper on cosmology in 1917. This work was remarkable in that it came less than a year after the landmark summary paper on general relativity; and, sandwiched between these monumental papers was, no less, the book explaining relativity to the layperson.[2] It seems that those years of endless tensor calculations that culminated in the 1915 breakthrough, rather than exhausting his creative energy (if I may use such a shadowy concept) – on the contrary, the tedious work apparently propelled him further. Perhaps mental inertia was at work, since the argument in his paper was a logical extension of general relativity.

Logical it was: although neither obvious nor inevitable. The title alone betrayed the logic: "Cosmological Considerations in the General Theory of Relativity."[3] Like the 1916 paper, the cosmology paper too involved a sequence of dense tensor calculus equations. Interestingly and uncharacteristically, near the start of the paper, Einstein injected this personal comment: "I shall conduct the reader over the road that I have myself travelled, rather a rough and winding road …." Fortunately, despite the almost impenetrable mathematics, as with general relativity, the physical meaning of the cosmology paper is quite simple and visually easy to grasp with the help of some further analog thinking. Avoiding, therefore, the mathematics of that rough and winding road, I hope the following explains clearly and correctly the essence of Einstein's cosmological considerations.

---

[1] As seen in Chap. 18, a few astronomers were testing the general theory, although it is questionable whether they understood the perplexing mathematics supporting the theory. We will return to this point in Chap. 23.

[2] Einstein [47] (1917).

[3] *Kosmologische Betractungen zur allgemeinen Relativitätstheorie*, in *Einstein Papers*, Vol. 6, Doc. 43.

D.R. Topper, *How Einstein Created Relativity out of Physics and Astronomy*, Astrophysics and Space Science Library 394, DOI 10.1007/978-1-4614-4782-5_21,
© Springer Science+Business Media New York 2013

General relativity explained gravity as being caused not by some spooky occult power acting instantaneously across space but as a manifestation of the local curvature of space around matter. The further question Einstein posed in this 1917 paper, as well as provided an answer for, is this: what happens if we sum-up these local curvatures of space across the entire universe? That is, how does the total matter of the universe affect the entire space? He found that the summation of all the local warpings of space by clumps of matter resulted in the bending of all space by all matter so that the entire universe was no longer infinite. The total space was, in fact, bent back within itself into a finite world. Einstein's universe therefore was, like the ancient cosmos, finite in space and matter. This finite world, however, was not bounded by a sphere of stars as was the ancient cosmos. On the contrary, like the Newtonian cosmos that Einstein was taught in school, it had no bounds – but with a difference. In the infinite universe, a traveler moving in a straight line would voyage forever without reaching an end; in Einstein's universe, travelers also never arrived at a boundary, except that in a finite amount of time they returned to where they began, since space was finite. It is important to point out that even though travelers returned to their starting point, they did not do so by turning around. Just as in infinite space, they voyaged continually in a straight line in one direction. How then could they return to their starting point in a finite time?

To grasp the answer requires first recalling the basic definition of a straight line – namely, the shortest distance between two points. Consider the following model as an analog, one that incidentally is right under our feet: the spherical Earth, with two poles, an equator, and lines of longitude and latitude. Along any line of longitude or along the equator, the line itself is the shortest distance between two points because all these lines are great circles, whose centers are the center of the Earth. (Lines of latitude are not, except for the equator.) It follows that any arc of a great circle is the shortest distance between two points. Locally such lines surely are straight, but the definition also implies that all arcs are therefore straight lines on the Earth's surface, even if the surface is curved.

This visualization is then applied to the two-dimensional analog from the last chapter: our two-dimensional person previously on an elastic flat space is now placed on the surface of a sphere (Fig. 21.1). Traveling in a straight line in any direction takes her along the arc of a great circle and she eventually arrives back where she started. She experiences this trip as moving away from her home base, going forward in a straight line, and yet, in time, coming back to base, without turning around. Mentally transposing this experience into our three-dimensional world, it means that space is not only locally curved around individual matter but is curved into the fourth-dimension as a whole. The total universe is warped so that if we set our spaceship to follow a straight line, and we hold that course, we will, eventually, return to where we started in a finite amount of time, the amount itself being a function of the size of the universe. Einstein's universe therefore was finite and unbounded, a sort of hybrid between the ancient (finite and bounded) and the Newtonian (infinite and unbounded) models. When those complex tensor calculus equations were transformed from mathematical formulae into a picture, it looked sort of like (as an analog) Fig. 21.1. The transformation, however and unfortunately, came with a deep puzzle.

Fig. 21.1 Einstein's
cosmological model of the
universe. The 2-D person
moving in a straight line in
her space returns to the
point of origin in a finite
time. From a 3-D viewpoint,
she is moving along the arc
of a great circle

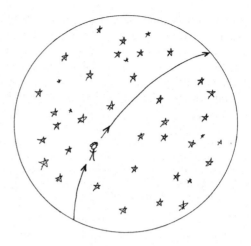

Both the ancient and the Newtonian models were static or stable. The stars on the ancient model were fixed on the stellar sphere, and, except for minor motions of individual stars, the Newtonian model too pictured an overall stability to the universe. In his cosmology paper Einstein pointed to "the fact of the small velocities of the stars" which implied a "quasi-static distribution of matter." The emphasis here was on the "small" and the "static." The "rough and winding road" of mathematics, however, predicted a much more unstable model; therefore Einstein believed it was "necessary" to stabilize it. Recall that Newton's initial universe likewise had a potential instability problem, in that it was prone to collapsing, as the mutual gravitational attraction of all matter coagulated potentially into one clump. To prevent this, Newton said that God "has placed [the stars] … at immense distances from one another." Einstein was faced with a similar problem. Although I am simplifying the mathematical complexity of the problem, in essence his finite universe was also prone to a similar gravitational collapse, and to prevent such an implosion he modified the deduced equation. He did this by introducing into the mathematics "a universal constant … at present unknown," which he labeled with the Greek letter lambda (λ), and later called the cosmological constant. As Duerbeck and Seitter write: "The cosmological constant counteracts the gravitational attraction of the masses; it avoids a gravitational collapse of the universe."[4] Shades of Newton's problem? Indeed, the constant stabilized an otherwise seemingly collapsing universe by introducing a sort of repulsive pressure throughout the finite universe, repelling the opposite gravitational attraction.[5] Moreover, by calling lambda "universal," Einstein implied that it was potentially on par with other

---

[4]Duerbeck and Seitter [34], p. 233. Their important article provides a detailed and technical history of cosmology from 1917 into the 1930s, with a larger cast of characters. This is a significant supplement to the overview of most popular and quasi-popular accounts, such as this book.

[5] It was later pointed-out by Eddington that the balance between these two forces was actually unstable: any perturbing of the system would result either in a total collapse or a continuing expansion.

constants of nature, such as the speed of light – and we know how he turned that constant into an absolute entity within special relativity.

The cosmology paper was published in February 1917; in the same month he wrote to Ehrenfest, as quoted in the epigraph commencing Part IV: "I have perpetuated something again … in gravitation theory, which exposes me a bit to the danger of being committed to a madhouse." In March he spoke of his paper this way:

> From the standpoint of astronomy, of course, I have erected but a lofty castle in the air. For me, though, it was a burning question whether the relativity concept can be followed through to the finish or whether it leads to contradictions. I am satisfied now that I was able to think the idea through to completion without encountering contradictions. Now I am no longer plagued with the problem, while previously it gave me no peace. Whether the model I formed for myself corresponds to reality is another question, about which we shall probably never gain information.[6]

The triumph of the achievement was thus tempered by some misgivings about the "reality" of the model. A possible source of this doubt is found in the following comment in a paper published in 1919:

> As I have shown in the [cosmology] … paper, the general theory of relativity requires that the universe be spatially finite. But this view of the universe necessitated … the introduction of a new universal constant λ, standing in a fixed relation to the total mass of the universe …. This is gravely detrimental to the formal beauty of the theory.[7]

An aesthetic matter cropped-up: he perceived a lack formal beauty in his cosmological theory – specifically in the ad hoc nature of the cosmological constant, which did not emerge naturally from the theoretical structure of the model – the sort of arbitrary entity that would gnaw at his being. An aesthetic problem vexing Einstein: where have we heard that before?

***

21.1   Summary

Einstein's 1917 cosmology paper begins with the general relativity deduction that gravitational force is caused by the local curvature, bending, or warping of space around every body of matter. He then sums-up the total universe of space and matter and deduces that the universal summation bends or curves the entire universe into itself, so that the resulting space is of finite extension. Einstein's cosmos is finite in extension but unbounded, in that travelers moving in straight lines (defined as the shortest distance between two points) arrive back at their starting points in a finite time. The two-dimensional analog would be a two-dimensional person living

[6]Letter to Willem de Sitter, March 1917: *Einstein Papers*, Vol. 8, Doc. 311.

[7]"Do Gravitational Fields Play an Essential Part in the Structure of the Elementary Particles of Matter?," pp. 191–198, in Einstein et al. [57] [1919], quotation on p. 193.

on the surface of a sphere. Furthermore, the equations of gravity in the paper implied an unstable universe, similar to the problem of gravitational collapse that Newton pondered. Einstein consequently stabilized his model by introducing a term (symbolized as λ), later called the cosmological constant, into the equation to repel or balance gravity. The constant appeared to be physically necessary, but aesthetically objectionable.

***

As seen in Chap. 13, Einstein's specific interpretation of what he called Mach's principle was a major factor in the development of general relativity. His fascination with the principle went back to his student days. The concept is found throughout his struggle with extending relativity to non-inertial systems, but the actual term "Mach's principle" only appeared in a paper of 1918, a year after the cosmology paper.[8] In this paper he explicitly separated this idea about inertia and the mutual interaction of matter from the relativity principle (that all motion is relative) and the equivalence principle. He saw these three, now distinct, ideas (a sort of conceptual trilogy) as sustaining the general theory. In a footnote he remarked that previously he linked the relativity principle with Mach's principle, and this "was confusing." Now Mach's principle is solely about explaining inertia by the interaction of bodies of matter. Further, he acknowledged that this obsession with the principle was mainly his alone: the "necessity to uphold" Mach's principle, he wrote, "is by no means shared by all colleagues; but I myself feel it is absolutely necessarily to satisfy it."[9] I previously spoke of Einstein in his earlier years being seduced by Mach's idea. Clearly, he still was.

The principle, it turned out, played a key role in his cosmology too, specifically the postulation of a finite (rather than an infinite) and a static universe. I hope this may be understood, without the technical mathematical apparatus otherwise necessary, by the following argument. In a finite cosmos there is the important realm of fixed stars at a finite extent; these stars are the "other matter" required for the mutual interaction essential in Mach's principle – a realm of matter necessary for both the analogy with gravity (about forces between two material objects) and the resulting explanation of inertia. Remember that the initial development of this model took place before the 1920s and Hubble's confirmation of external galaxies. Therefore an infinite universe for Einstein would probably consist of our Milky Way, held together

---

[8]*Einstein Papers*, Vol. 7, Doc. 4. The title of the paper is *Prinzipielles zur Allgemeinen Relativitätstheorie*, or "Fundamentals to the General Theory of Relativity." The editors of the *Papers* strangely translate this as "On the Foundations of the General Theory of Relativity," pp. 33–35 ET, which is nearly the identical translation used for the 1916 landmark summary paper, *Einstein Papers*, Vol. 6, Doc 30, *Die Grundlage der Allgemeinen Relativitätstheorie*, or "The Foundation of the General Theory of Relativity," pp. 146–200 ET. This results in an unnecessary confusion between the two distinct papers, and I am perplexed as to why they did this, since the German titles were indeed different.

[9]*Einstein Papers*, Vol. 7, Doc. 4, pp. 33–35 ET.

by gravity, at the "center" of an infinite space. By a process analogous to the evaporation of the molecules of a gas, matter (that is, the stars) in our Milky Way over time would slowly dissipate throughout this universe.[10] This conception would contradict the possibility of an infinite universe lasting very long. In the finite cosmos, however, there was a balance between gravity attracting matter by the inverse-square law and the cosmic repulsion – namely, the cosmological constant, lambda (λ) – acting directly-proportional to the distance between all bodies of matter.[11] This cosmos remained static and stable by a balance between these two forces. For these reasons the universe was finite, not infinite, and Mach's principle remained intact.

Einstein, however, would come to abandon this viewpoint, specifically after meeting Hubble, as will be seen near the end of Chap. 23, below.

---

[10] Kragh [122], p. 131.

[11] I mention the direct proportionality of this law, as distinct from the inverse-square law of gravity, but am not pursuing the mathematics of this any further.

Chapter 22
Three Challenges to Einstein's Cosmic Model

Einstein's introduction of the cosmological constant was seen by later scientists as at least unnecessary and at most a mistake. Einstein himself eventually called it the "biggest blunder of my life."[1] More significantly, the important historical question is this: was Einstein's introduction of λ a reasonable assumption at the time? I think the answer is "yes" from what we know of his astronomical information in 1917. As seen, in the early twentieth century, there was growing evidence of major motions of matter in the universe, pointing toward a modification or revision of its structure. Yet as stressed above, Einstein was not an astronomer; furthermore it is important to keep in mind the insularity among different branches or communities of physics then, as now. Physicists cannot keep up-to-date with frontier knowledge in all divisions of physics; one or two is enough for a lifetime. In Einstein's time, just as most observational astronomers had little knowledge of general relativity, with few being even minimally familiar with tensor calculus, the small theoretical physics community was thoroughly absorbed with matters around relativity and the newly discovered subatomic quantum world, which left little time for keeping up with details of what the big telescopes were finding as they probed deep space. A fair-minded conclusion is that Einstein's introduction of the cosmological constant was a reasonable assertion for him in 1917.

---

[1]This comment was attributed to Einstein by the Russian-American physicist, cosmologist, and popularizer of science, George Gamow. Gamow did not date the comment but just said it was "much later" than the early 1920s. The exact quotation is that Einstein "remarked that the introduction of the cosmological term was the biggest blunder he ever made in his life" [78], p. 44. Gamow first mentioned it in an article in September of 1956 in *Scientific American*, which he cited in the bibliography to his 1970 autobiography. Gamow's statement was reinforced by the American cosmologist Ralph A. Alpher, who visited Einstein with Gamow "about 1952," he recalled. Alpher further remembered that Einstein said his "introduction of the concept [of the cosmological constant] in his early work was a blunder." Alpher was remembering this in 1998. If Alpher was correct, then the remark was from the early 1950s. Quoted from an email [3]. Also present at the meeting was Robert Herman. For more on Alpher and Herman, see the discussion of the discovery of the cosmic background radiation in Chap. 24.

D.R. Topper, *How Einstein Created Relativity out of Physics and Astronomy*, Astrophysics   165
and Space Science Library 394, DOI 10.1007/978-1-4614-4782-5_22,
© Springer Science+Business Media New York 2013

The next decade, however, was a different story. It began with the Dutch astronomer, Willem de Sitter, who met Einstein in Leiden in the fall of 1916, and who was one of the few astronomers also proficient in manipulating the mathematics of general relativity. At the time of publication of Einstein's cosmology paper, the two were already corresponding over matters pertaining to general relativity, some of which involved de Sitter's skepticism over Mach's principle and Einstein's defense of it.[2]

After de Sitter read Einstein's cosmology paper the subject shifted to that topic. In fact, I quoted a few a pages ago from a letter to de Sitter in March 1917, where Einstein called his theory "a lofty castle in the air," and further questioned if "the model I formed for myself corresponds to reality," while seemingly concluding that "we shall probably never gain [such] information."[3] Was Einstein truly this skeptical about his model?

De Sitter was; especially about the necessity of the cosmological term, and he put forth the possibility of avoiding it by making other assumptions about the universe. Since so much of the universe is empty space, with matter scattered throughout, he applied Einstein's cosmological equation, without the λ, to a matter-free universe, and found that it was stable. De Sitter's empty model had another property; it produced the illusion of the stars receding from us. Being an astronomer, he was knowledgeable of Slipher's work, and so he pointed to the redshift of most nebulae as possible empirical support for the apparent recession implied in his model. So de Sitter wrote Einstein on this; and, perhaps predictably, Einstein summarily rejected de Sitter's model.[4] A recession of stars, real or apparent, he believed contradicted his stability assumption; and even more adversely, an essentially empty universe had no correlation to reality. De Sitter wrote back in defense but Einstein was adamant. "In my opinion, it would be unsatisfactory if a world without matter were possible."[5] The contrast between these viewpoints was cleverly characterized by Eddington this way: Einstein put forth a world of matter without motion, and de Sitter a world of motion without matter (see Photo 22.1).[6] At the time de Sitter's was the first of three challenges to Einstein in the decade following the 1917 paper.[7]

The second came from the Russian-born mathematician Aleksandr Friedmann in two papers (1922 and 1924). Coming from a mathematician, Friedmann's objections were directed toward the internal mathematics of Einstein's paper, not explicitly the physical application. His results, however, did show that by eliminating the cosmological constant the equation revealed a logically possible model, albeit still

---

[2] *Einstein Papers*, Vol. 8, Docs. 272 and 273. There were over 20 exchanges between them from June 1916 to April 1918. See also Kahn and Kahn [112].

[3] *Einstein Papers*, Vol. 8, Doc. 311.

[4] Topper [198], p. 196; Bartusiak [8], pp. 142–145.

[5] *Einstein papers*, Vol. 8, Doc. 317.

[6] Bartusiak [8], p. 143.

[7] For a thorough analysis of their exchange of ideas, with mathematical details far beyond the scope of this paper, see the excellent paper of Realdi and Peruzzi [170], and Kerszberg's book [115].

Photo 22.1 Einstein at the Leiden Observatory with, front row, Arthur S. Eddington and Hendrik A. Lorentz; back row: Einstein, Paul Ehrenfest, and Willem de Sitter, September 1923. Permission AIP Emilio Segre Visual Archives

unstable. Einstein again rejected the implication, remaining committed to a stable physical model.[8]

In 1927 the Belgian physicist and Jesuit priest Georges Lemaître,[9] unaware of Friedmann's work, also eliminated Einstein's constant and showed that the resulting model implied a continually expanding universe, an expansion that began from Einstein's initial static model.[10] The English title of his paper (published in French) summarizes the idea: "A Homogeneous Universe of Constant Mass and Increasing Radius Accounting for the Radial Velocities of Extra-Galactic Nebulae."

As a student he had written a thesis on relativity and gravity which won him a traveling scholarship from the Belgium government. Lemaître first studied in England at Cambridge (under Eddington, focusing on relativity), and later traveled to the United States, studying at Harvard with Shapley, and then obtaining a second Ph.D. at MIT.[11] In 1925 he was present at the reading of Hubble's paper at the AAS meeting;

[8] Topper [198], p. 196. Kragh and Smith [123], pp. 145–147. Sadly, not long after this exchange Friedmann died.

[9] (1894–1966).

[10] Kragh and Smith [123], p. 146.

[11] Kragh [122], p. 143; Farrell [61], p. 90.

he then traveled to the Lowell Observatory in Arizona to visit Slipher, follow by a jaunt to California to met Hubble. This, in part, explains his pointing to empirical support for his theory from observational astronomy, since he was one of the few physicists up-to-date on the subject, having personally met the key players.[12]

About six months after the publication of his paper, Lemaître met Einstein for the first time at a conference in Brussels. Einstein was not aware of Lemaître's paper, for (as noted) it was published in French, and in an obscure journal. Indeed, the paper was mostly ignored by the scientific community. Upon listening to Lemaître explain his theory Einstein mentioned Friedmann's work, which Lemaître had not heard of. It appears that Einstein then mentally lumped Lemaître's idea with Friedmann's, by summarily rejecting them both. Stubbornly holding to his commitment to a static universe in equilibrium, Einstein is reported as saying to him, "Your calculations are correct, but your physical insight is abominable."[13] Dismissed by Einstein, Lemaître's paper languished in cosmological limbo until being recognized by Eddington, who reprinted it in the <u>Monthly Notices of the Royal Astronomical Society</u> in 1931.[14] By then (as shall be seen), the cosmological milieu had radically changed, mainly because of the further work of Hubble.

Einstein's striking-out, one-two-three, from de Sitter, through Friedmann, to Lemaître may seem unreasonable from today's viewpoint. It is, needless to say, easy to perceive previous mistakes when the correct answer is known. Try, however, to see the issue from Einstein's perspective. De Sitter's model without matter had little physical meaning to him. Similarly he saw Friedmann's effort as purely mathematical manipulations. It would take more physical insights to convince him otherwise. After all, Einstein was a theoretical physicist, with the emphasis on physics. Recall his initial reluctance to take Minkowski's extra-dimensional interpretation of special relativity seriously. Even those subsequent years of grinding away at tensor calculus did not compel him to loosen his link with the physical world behind the mathematics. Recall too what he said later in his autobiography: there is a "huge world, which exists independently of us human beings" – a physical world, albeit describable by mathematics, but essentially an independent "extra-personal world" nonetheless.[15]

---

[12]McVittie [140].

[13]Quoted in Topper [198], pp. 196–197; found quoted in Smith [185], p. 57.

[14]Lemaître [129] [1927]; McVittie [140]. The English translation deleted some of the text and most of the footnotes: see Kragh [121], p. 406 n 28. Further details of the changes in the translation were the focus of an article by Van den Bergh 2011, where he lamented that the identity of the translator and the reasons for the deletions remained unknown. There were even intimations by others of a conspiracy by Hubble to get further credit for the expanding universe. But the puzzle was recently solved by Mario Livio [133], who discovered that the translator was indeed Lemaître himself (!), and he purposely omitted some passages. Clearly Lemaître was not obsessed with matters of priority but with eliminating material he thought was out of date. Livio found the evidence among letters in the Lemaître Archives and the Royal Astronomical Society correspondence and in the minutes of their meetings.

[15]Einstein [51] (1949), p. 5.

How do we explain his dismissal of Lemaître model? On the one hand, it was not easy for Einstein to abandon the cosmological constant, since a stable universe was a priori taken for granted. Even if one assumed stars or nebulae receding from us in all directions, as implied by Slipher's work on redshift, this motion was usually thought of as chunks of matter moving within empty space; and this space was still the Euclidean space of Newton. That same expansion, however, when applied to Einstein's finite universe, implied something else – something quite different and extraordinary. Using the two-dimensional analog (Fig. 21.1), replace the stars with nebulae (in a post-Hubble world), and impart to the sphere some elasticity, rather like a balloon. Blowing up the balloon would correspond to an expanding universe; from the point of view at any place on the sphere, all the stars or nebulae would be seen as moving away. Since the surface of the balloon is space itself then as the balloon grows, space is being created as the matter moves away within the expanding (non-Euclidean) space. The movement of matter, therefore, is not through space, but rather matter is carried along within the expanding space itself. Time and (non-Euclidean) space are continually come into existence, drawing along matter with them. Einstein – who had relativized time in special relativity and locally warped space in general relativity – was not yet ready to go this far. Even he, it seems, had some limits to his imagination.

What, however, to do with Lemaître pointing to empirical support from astronomy? It appears that Einstein as yet did not appreciate the discoveries coming from observational astronomy; he had not fully absorbed their significance. To him, Lemaître was making a large and unreasonable leap from Slipher's data of moving nebulae. But all that would change when Einstein met Hubble in 1931. To set the stage for that story we need to bring our survey of observational astronomy up to that time.

Chapter 23
1931: Caltech, Again; Einstein Meets Hubble

Hubble's announcement in 1925 of nebulae external to our Milky Way was momentous. Those few puzzling blurs in the night sky were mostly ignored over the millennia of astronomical history, as the focus instead was on the Sun and Moon, planets and stars. But the nebulae, it turned out, were ultimately the essential structures in the universe. The historical process, not unexpectedly, went slowly. There was a reluctance to accept the nebulae as being external to our galaxy home; once accepted as being external, there was still a residual homocentric hesitation to conceive of them as being very far and very large, especially larger than our own Milky Way. The aversion to abandoning once again our centrality and uniqueness can be gauged by the evolution of terminology: the external nebulae were sometimes called nongalactic nebulae, cosmic nebulae, extra-galactic nebulae, and, of course, island universes. The term galaxy was too synonymous with our Milky Way to give up easily. Hubble, when he died in 1953, still called them nebulae. The issue, I submit was similar to the difficulty in post-Copernican times of calling our Earth a planet.

Hubble's discovery, however, was just the first stage of what in time became a two-staged breakthrough. In this second stage he worked with his assistant, Milton Humason, who developed extraordinary skills at photographing spectra, despite his more modest beginnings as a mule driver with an eighth grade education hauling equipment up Mount Wilson. (Incidentally, Humason seems to be the assistant in the tale about Shapley rubbing out Cepheid variables in the photo of the Andromeda nebula.[1]) Hubble and Humason picked-up where Slipher left off (in fact, they initially borrowed his data for many nebulae without citing him in their first publication on the topic). Slipher had reached a limit using the telescope at the Lowell Observatory, whereas Hubble and Humason employed the most powerful telescope in the world at Mt. Wilson, which was able to penetrate deeper into space, and they went on to measure spectral shifts of more external nebulae. They found that most

---

[1]For an in-depth probe into the likelihood of this story, and possible ramifications on the subsequent careers of Shapley and Hubble, see Sandage [177], pp. 495–498.

D.R. Topper, *How Einstein Created Relativity out of Physics and Astronomy*, Astrophysics and Space Science Library 394, DOI 10.1007/978-1-4614-4782-5_23,
© Springer Science+Business Media New York 2013

Fig. 23.1 Hubble and
Humason's linear
correlation between the
redshifts and the distances
of nebulae

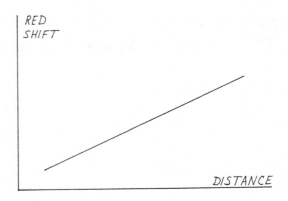

of the nebulae exhibited redshift. Incidentally, the catalyst for Hubble's redshift pursuit possibly was a chat with de Sitter at an international scientific meeting in Holland in the summer of 1928, when de Sitter encouraged him to test the expanding hypothesis.[2] Subsequently, on arriving back in California, Hubble soon commenced the redshift quest. Otherwise, it seems that Hubble had no knowledge of Friedmann's or of Lemaître's work, and therefore had no reason to pursue the redshifts of the nebulae.[3]

He and Humason found not only redshifts for more nebulae but the nebulae were distributed in all directions. If these redshifts were indeed Doppler shifts, then this appeared to support de Sitter's model (if not Lemaître's, too). Hubble the empiricist, however, resisted such speculation, although he did link the redshift to receding velocities. More important to him was something else he discovered: measuring the distances of the nebulae he found a correlation between these distances and the amount of redshift. When graphed, the result was a linear relationship as in Fig. 23.1. He and Humason published such a graph for several dozen nebulae (many borrowed from Slipher) in a landmark paper in March, 1929. The vertical axis is the redshift, which they labeled velocity, retaining Slipher's identification. Yet, notably, in the text they speak of "apparent" velocities, and only at the very end of the paper do they mention the "possibility" of a connection of this graph with de Sitter's model. In the spring of 1931, twenty-six months later, in a second paper, they further confirmed the linear relationship for forty more nebulae.[4] Today this graph is called Hubble's law and is a fundamental component of astronomy. At the time, however, it was a new puzzle that arose as astronomers were still adjusting to the apparent demotion of our Milky Way to being just another galaxy.

Hubble interpreted the graph primarily as an empirical correlation employed to measure celestial distances beyond what can be achieved with Leavitt's law, which

---

[2]Bartusiak [8], p. 226.

[3]Sandage [177], p. 502.

[4]Bartusiak [8], p. 233; Sandage [177], pp. 502–507.

requires the detection of Cepheid variables within distant nebulae. Most nebulae are too far away to see individual stars, but their redshift is still visible. Assuming a continuing linear relationship beyond the nebulae whose distances are measured, the graph delivers a straight-forward method of deriving the extreme distances of nebulae by measuring their redshift alone. In short, Hubble's correlation extended the range of measuring cosmic distances far beyond the limit of Levitt's law. This was the core of Hubble's discovery to him; he was not prone to speculate further about hypothetical cosmological models.

Hubble of course was free to eschew the physical interpretations of his graph; nonetheless, if redshifts were Doppler shifts, this meant that the nebulae were mostly receding from us. Could it be otherwise? One answer came from the respected astronomer Fritz Zwicky[5] at Caltech. Conceding that redshifts are Doppler shifts for local nebulae, Zwicky questioned applying the relationship to the total universe. Drawing on Einstein's discovery that light is bent by gravity, Zwicky argued that light-speed may be retarded as it travels through space over eons of time, and especially as it recurrently passes and is bent by large masses. He called this slowing-down of light-speed the "gravitational drag of light" (others later named it the "tired-light" hypothesis). This drag on light would, in turn, produce redshifts in their wavelength, independently of their possible recessional motion. Redshifts, therefore, were not necessarily caused by motion, and thus no excessive recession of nebulae (or galaxies) was required or assumed to explain them.[6] This was one way of solving the puzzle.

Such matters were brewing within the astronomical community when Einstein and his entourage arrived in California – a center of cosmological conjecture and controversy, and hard empirical data too.

* * *

As seen near the end of the last chapter, during Einstein's first winter sojourn at Caltech (from late December 1930 to early March 1931) he conversed with several key experimental scientists who had or were testing relativity. On the cosmological deductions of the theory, the key person was Hubble along with Humason, who in March 1929 first published the discovery of the redshift-distance correlation and, not long after Einstein left California, they published the second further confirmation in the spring of 1931. It was, no doubt, an auspicious meeting for all. Regrettably, there seems to be little direct documentation available of the details of what happened, and, surprisingly, Einstein recounts in his diary only mundane matters about the trip.[7]

Fortunately, due to Einstein's celebrity status, The New York Times sent a reporter to transmit daily dispatches about his every move. The reporter not only spied on Einstein's eating habits at restaurants, but happily sat in on an occasional

---

[5] (1898–1974). A Swiss-Czech astronomer who was educated at the ETH when Einstein was teaching there. Marianoff [135], p.145, says Zwicky studied under Einstein. Zwicky immigrated to the United States in 1925, and spent the rest of his scientific career in Caltech.

[6] Topper [198], pp. 78–79.

[7] I am grateful to the late Martin J. Klein for a copy of parts of Einstein's travel diary, which I obtained when Klein was director of the Einstein Papers Project.

lecture. Here is the first important quotation from Einstein, reported on January 3, 1931 on the front page of the newspaper, from a lecture deliver the previous day: "New observations by Hubble and Humason ... concerning the redshift of light in distant nebulae make the presumption near [that is, make it appear likely] that the general structure of the universe is not static." This is the initial inkling that Einstein was budging from his previous stubborn resistance to anything but a static universe. After extensive meetings with Hubble and Humason, according to another front page story on February 5, Einstein announced at a lecture on the previous day that he no longer held to the model of a stable universe. Then in a lecture on February 11, Einstein was quoted the next day as confessing: "The redshift of distant nebulae has smashed my old construction [or, model] like a hammer blow," and he was further reported as "swinging down his hand to illustrate" the obliteration of his static cosmic model, as he professed his change of opinion.[8] Two months with Hubble were enough to pry him loose from his attachment to the cosmological constant. By March, when he left California to the open spaces of a visit to the Grand Canyon, he realized that the postulation of that constant was a major blunder.[9]

A puzzle worth contemplating here is the difference between Hubble's and Einstein's interpretations of the redshift of the nebulae. The documented evidence, albeit scant, is that Einstein's exposure to Hubble's work made him rather quickly change his mind on the stability of the cosmic model. Most probably, Hubble shared with Einstein his work with Humason not only from the 1929 paper but also on the important 1931 paper that was coming out in the spring. But what about Hubble's reticence to recognize redshifts as Doppler shifts? One would think that this attitude would have been communicated to Einstein during their chats together. So why didn't Hubble's misgiving rub-off, reinforcing Einstein's refusal to go along with the unstable universe? Indeed, given Einstein's previous dogged resistance to giving up his stability postulate, it is rather astonishing that he so easily changed his mind, after balking at the suggestions to the contrary for over a decade.

Part of an answer may go back to 1905, where, as pointed out before, he also deduced an optical Doppler effect from his electromagnetic revision of Maxwell's equations. Accepting the Doppler effect for light meant that the redshifts were measuring receding nebulae. Because of this, the cosmological constant had to go.

There is another matter to confront at this juncture. Einstein's abandonment of the closed universe necessitated another significant abandonment: namely, Mach's principle. For when Einstein acknowledged that the universe was not stable and that an expansion was real, he realized he had to reject Mach's principle.[10] The argument seemed to be as follows: If matter was indeed receding, then there was no fixed "other" matter out-there to provide the mutual interaction, as the fixed stars did, which was

---

[8] Quoted in Topper [198], pp. 197–198; and quoted in Topper and Vincent [199], 281–282.

[9] In the 1930s Einstein added an appendix (IV) to his popular book on relativity, conceding the roles of Friedman and Hubble in convincing him of the expanding universe. Einstein [49], pp. 133–134. Curiously he does not mention de Sitter or Lemaître.

[10] Theorists today debate whether this rejection was necessary (see, e.g., Hoefer [96], pp. 87–90), but this issue is irrelevant here, since we are only interested in what Einstein believed.

essential for explaining inertia without absolute space. An expanding universe, therefore, was not compatible with Mach's principle. This certainly was a key source of his
rejection, at least by the 1930s, of the principle that had obsessed him for so many
years. This also meant that there was no relativistic explanation of inertia.

What a significant visit this trip to Caltech was: Einstein summarily rejected two
previous deeply held notions (the cosmological constant and Mach's principle)
when confronted with experimental evidence to the contrary, showing further the
empirical side of him.

But there was another reason to doubt the certitude of Mach's principle. Something
else was brewing in the 1920s as result of his attempt at a unified field theory, which
is the topic of the next (and last) Part of this book. He (and therefore we) will confront Mach one more time.

* * *

Having spent the winter of 1930–31 "loafing" in California, Einstein returned to
Caltech the next two winters. As if the scenarios were conjured by malevolent gods,
he met de Sitter and Lemaître in those subsequent years, they being two of the three
theoreticians who had challenged his cosmological constant in the 1920s. Friedmann,
one surmises, the gods would have thrown in too, if he had not died.

In the second winter (1931–32) de Sitter was visiting Caltech and interestingly
he and Einstein wrote a paper (published in 1932) on a cosmic model using a flat
space without the cosmological constant. Pierre Kerszberg quotes Eddington as
reporting that Einstein told him that the paper was not important but de Sitter though
it was; whereas de Sitter told Eddington that he thought the paper was of little
importance but "Einstein seemed to think it was [important!]."[11] This leaves us with
an unclear assessment of the meeting.[12] It was the last of the Einstein–de Sitter col-
laborations, since de Sitter died in 1934.[13]

During the next winter term (1932–33) he met Lemaître, who was lecturing at
Caltech. Since his 1927 paper on an expanding universe, Lemaître was now speculating on the initial state of the universe, speaking of it as a "primeval atom."[14]
Einstein attended the lectures but his response is uncertain since the documentation
is ambiguous. On the one hand, Einstein is quoted as saying: "This is the most beautiful and satisfactory explanation of creation to which I have ever listened."[15] On the
other hand, the remark may not have been directed to the explanation of creation,
for another source affirms that Einstein called the lecture a "beautiful and satisfactory explanation of cosmic rays."[16] Why the confusion? We do not know precisely

---

[11] Kerszberg [115], p. 361n.

[12] Kragh [121], p. 35, incidentally, notes the importance of the model in later cosmology.

[13] Pais [162], p. 494.

[14] He first presented this in 1931 at a meeting of the British Association for the Advancement of
Science.

[15] Quoted in Kragh [121], p. 55. Also quoted in Michelmore [141], p. 176.

[16] Kragh [121], p.55, especially note 93 on p. 408.

why, but part of answer may be gleaned from the conceptual relationship at the time between cosmic rays and the idea of a creation. Today we know that so-called cosmic rays are high-energy subatomic particles (such as protons) coming from the Sun, other stars, and elsewhere, some still of unknown specific origins. The term was coined by Millikan, since he incorrectly believed they were high-energy photons (hence light "rays"); he was right, however, in proposing that their origin was extraterrestrial ("cosmic"). Beyond his administrative duties at Caltech, Millikan's specific area of scientific inquiry was the study of, and speculation about, cosmic rays. He thought they came from newly created atoms, essentially God's way of preventing the universe from running down. In his lecture, however, Lemaître had proposed a slightly different source: he argued that cosmic rays were leftover stuff[17] from the early universe, thus supporting his primeval atom hypothesis.[18]

Which of the two hypotheses did Einstein find beautiful? Ultimately we do not know. But the following may cast some light on the matter. In 1958 Lemaître recalled their meeting at Caltech, and wrote:

> As I spoke with him [i.e., Einstein] about my ideas regarding the origin of cosmic rays, he said excitingly, "Have you spoken with Millikan?" [B]ut when I spoke to him about the Primeval Atom, he interrupted me, "No, not that, that suggests too much the creation."[19]

This remark would suggest that Einstein accepted Milliken's origin of the rays but not that of Lemaître. John Farrell, who quotes the above recollection, is justifiably puzzled with Einstein's remark; as he points out, Einstein later was willing to write of a beginning to the universe.[20] Farrell specifically refers to the 1945 appendix to the second edition of the Princeton Lectures.[21] There Einstein put forward the rejection of the cosmological constant and the subsequent acceptance of an expanding universe, from which it logically followed that there was a "beginning of the world" such that "the now existing stars and systems of stars ... did not yet exist as individual entities."[22] Farrell goes on, attempting a resolution of the contradiction by trying to read Einstein's mind:

> [Einstein] had enough philosophical grounding to realize that an origin of space-time was not the same thing as creation of the world out of nothing, a concept he appreciated was intrinsically outside scientific bounds.[23]

---

[17]Farrell says Lemaître spoke of "fires and smoke" but does not cite a source. Farrell [61], p. 101.

[18] McVittie says Lemaître interpreted the rays as "remnants of the original cataclysm." McVittie [140], p. 297.

[19]Quoted in Farrell [61], p. 100.

[20]Farrell [61], p. 102.

[21]Einstein [48] [1945], pp. 109–132.

[22]Einstein [48] [1945], p. 129.

[23]Farrell [61], p. 102.

This apparent separation of the scientific and theological is a reasonable assessment of what Einstein may have implied in his quick response to Lemaître, except that – to be seen in Chap. 28 – Einstein was not averse to speaking of God in a scientific context when it came to his disagreement with other scientists over their interpretation of quantum physics. So we come back to where we began: Einstein's response is intriguing and worth pondering, but the real meaning remains unknown.

Finally, this 1933 meeting produced little or no advance in Einstein's own work on his cosmic model; instead, others carried the topic forward.[24] For, by then, his focus of scientific attention had turned to a different topic – the subject of Part **V**, when we return one last time to Caltech in 1931 and an explanation of the equation on the blackboard in Fig. 18.1, Photo 18.1.

---

[24]For more details beyond the scope of this book, for the period from 1916 into especially the 1930s, see Kerszberg [115].

Chapter 24
Cosmology Since 1931: Highlights and Episodes

This chapter is a survey of cosmology from the formative juncture around 1931, when empirical astronomy met modern theory, and proceeds through some of the major discoveries, episodes, and highlights approaching the present. (The reader itching to follow Einstein's path, may hurdle to Chap. 25.)

What became known as Hubble's law (Fig. 23.1) was interpreted as visualizing an expanding universe, if the redshifts were Doppler shifts. As such, the slope of the straight line was correlated with the speed of expansion, and from this the age of the universe could be calculated, since the time variable is contained in the slope.[1] Using the 1931 data of Hubble and Humason, and applying this to the expanding model (think of the balloon analog), the age of the universe was calculated as about two-billion years. At the time this was conceived as a reasonably high number for astronomy, but there was a major problem. There was a parallel development in atomic and nuclear physics; from the growing knowledge of radioactivity and nuclear decay, a method of geological dating had been devised, which gave a result of three- to four-billion years for the Earth's age (today the age is about four-and-a-half-billion years). Since this implied that the Earth came about considerably after the universe began, there was a fundamental contradiction between two branches of physics. As Einstein wrote in an appendix to his popular book on relativity in the 1930s: "It is in no way known how this incongruity is to be overcome."[2] This contradiction gave credence to those questioning the Doppler interpretation of redshift, along with other doubts about the expanding model.

This apparent roadblock to the progress of cosmology virtually halted further work on the subject. (This may have been another reason for Einstein's dearth of

---

[1]The calculation is quite simple. If the vertical axis is speed, which is distance/time, and the horizontal axis is distance, then the slope is distance/time/distance/1, which reduces to 1/time. This was later written as $V = H \times d$, where v is velocity or speed, d is distance, and H became known as Hubble's constant. Using this notation, then $1/H$ is a measure of the age of the universe.

[2]Einstein [49], p. 134.

D.R. Topper, *How Einstein Created Relativity out of Physics and Astronomy*, Astrophysics and Space Science Library 394, DOI 10.1007/978-1-4614-4782-5_24, © Springer Science+Business Media New York 2013
179

work on cosmology.) Observational astronomy, however continued its progress, but theoretical speculation was minimal. Undoubtedly, theoretical physics (other than cosmology) had other more promising fields to sow: quantum physics and especially the subfield of nuclear physics. A noted before (Chap. 10), understanding the world of the nucleus resolved the ancient problem of why the Sun and stars don't burn-out. The nuclear process itself explained the life and death cycles of stars and variations thereof, which accounted for different types of stars. Astrophysics was making considerable progress, while cosmology languished.

An eventual revival of cosmology began around 1946 in, of all places, a movie theater, when three young Cambridge University astronomers attended a ghost movie titled <u>Dead of Night</u>. The plot involved five stories that were linked so that the beginning of the film became the end, and thus time was in a loop. Inspired by this film, or so the story is told, the three – Thomas Gold, Hermann Bondi, and Fred Hoyle – wrote a paper that put forward a cosmological model that avoided the contradiction with geological time.[3] This bold paper, published in 1948, postulated a universe with no beginning or end, where matter is continually but minutely (a few atoms per cubic mile) being created in the space left by the receding galaxies (recall the life-cycles of stars, mentioned above), and thus over eons of time nothing really changes. Named the steady state model, it eliminated the contradiction with the age of the Earth, since there was no creation. The lack of a beginning was also seen as eliminating a potential theological problem involving the need for a source (or cause) of the universe – although, in a sense, it boiled-down to two choices of creation: whether matter is continually being created, or that it happened all at once.

As the steady state model became increasingly adopted by astronomers in the early 1950s, the large telescopes in California (such as at Mt. Palomar) were producing data that led to a recalibration of Hubble's law and a corresponding change in the value of Hubble's constant, H (see footnote 1, above). The result was a decrease in the original slope of the line in Fig. 23.1, which meant that the time of expansion was extended. Applying this to the expanding model, the age of the universe was increased to about ten-billion years, and the contradiction with the Earth's age was no more. There was now an equal choice between the two alternate models. Yet, mainly for aesthetic and, in some cases, theological reasons, many cosmologists remained committed to the steady state model.

An interesting parallel topic is the belief of Hubble himself, who died on September 28, 1953. As noted before, in the 1930s he was skeptical as to the Doppler interpretation of the redshifts. On May 8, 1953 Hubble delivered the prestigious George Darwin lecture, which he titled, "The Law of Red-Shifts."[4] Clearly this was his last word on the subject. In the first sentence he defined the law as "the correlation between distances of nebulae and displacements in their spectra." Note two things (or lack thereof): the use of the word nebulae not galaxies, and the word spectra not velocity or speed. He did, nonetheless, speak of velocities in the paper

[3]Gold, Austrian (1920–2004). Bondi, Anglo-Austrian (1919–2005). Hoyle, English (1915–2001).
[4]Hubble [105].

(especially when referring to the work of Slipher), but what was striking throughout the paper was the cautionary tone regarding the redshifts as being Doppler shifts. At one point he said: "Regardless of the interpretation of red-shifts, we must accept the loss of energy by the individual quanta...," which harkens back to Zwicky's tired-light idea. Later he acknowledged that "if red-shifts are interpreted a Doppler shifts," then the age of the universe according to the expansion model is three- to four-billion years "and thus comparable with the age of rock in the crust of the Earth."[5] At the time of Hubble's lecture, he was using data from Palomar into 1951, and the expansion law had been recalibrated to deducing three- to four-billion years for the age of the universe, almost twice the 1930s number of two-billion, which he notes. As noted above, further into the early 1950s that number was extended to ten-billion, although the smaller range, at least, resolved the contradiction of the calculated age of the universe with age of the Earth.

Hubble, thus, died as that recalibration was in progress. Regardless, according to Allan Sandage,[6] Hubble never accepted the actual expansion of the universe.[7] Sandage was Hubble's assistant during his last years (1949–1953), and called him a "pragmatic observer." Sandage asserted that up to the end Hubble believed "that the redshifts were probably not true velocity shifts – and that an unknown law of nature must be invoked" to explain them.[8] This tallies with what was implied in Hubble's last lecture – that to the end he was the consummate pragmatist. Listen to what he said in his last paragraph:

> Today we have reached far into space. Our immediate neighbourhood we know rather intimately. But with increasing distance our knowledge fades, and fades rapidly, until at the last dim horizon we search among ghostly errors of observations for landmarks that are scarcely more substantial.[9]

A pragmatic observer, indeed, and an empirical skeptic to boot.

Throughout the 1950s there was a choice between two models, and an alternate name for what was called the expansion model. Hoyle, as a droll and disparaging joke, called it the big bang model, a term which unfortunately stuck.[10] The aim was to ridicule the need for a beginning, when all energy in the universe would be concentrated into one point – a problem precluded by the steady state model.

The 1960s were a transitional decade with three major episodes. In the early years (1960–64) radio telescopes from several observatories first detected point-like sources of enormous energy within radio-waves of the spectrum. Having so much

---

[5]Hubble [105], pp. 658, 664, and 666.

[6]American astronomer (1926–2010).

[7]From a personal communication with Sandage, May 1, 1998, who alerted me to Hubble's last lecture. Topper and Vincent [199], p. 288, note 21. Also Sandage [177], pp. 516–521, and Brush [19], pp. 123–125.

[8]Sandage [177], p. 502 and 518.

[9]Hubble [105], p. 666. Also see Kragh and Smith [123], pp. 149–152.

[10]For some reason, the term is usually capitalized, but I have not done so.

energy, they could not be stars, but they could not be galaxies either; not only because of the enormous concentration of energy into a small space, but because they changed in brightness over short periods of time (from weeks to months), and galaxies cannot pulse in this manner (recall that it takes 100,000 years for light to travel across our Milky Way). Initially, therefore, they were called quasi-stellar radio sources or quasi-stellar objects – star-like, but not real stars – and this was later shortened to the word quasar. Totally unexpected, with no hypothetical rational, they came on as a new class of objects in the sky – and as a mammoth puzzle. How could the energy of a galaxy be compacted into the space of star?

As further quasars were searched out (the large telescope at Palomar was of major importance in this task), more were found, with some being within the visible spectra. Yet more important was this: another piece was added to the puzzle when it was found that quasars exhibited redshift, at first about one-fourth the speed of light, and then approaching light-speed. What did that mean? Interpreting the redshifts as Doppler shifts put some quasars near the edge of the universe, which was also near the beginning of the universe on the expansion model. The corresponding question was: How can something so large (in energy) be so far away and still be visible? They seemed to defy the laws of physics; unless they were smaller and much closer than previously thought, and, as such, they could also pulse without breaking any physical laws. This alternative viewpoint was very reasonable, if the redshifts shifts were not Doppler shifts. In fact, this was compatible with the steady state model, which placed the quasars nearby and explained away the redshifts as due to internal forces such a gravity stretching the wavelengths of light toward the longer red side of the spectrum (shades of Einstein's gravitational redshift). By mid-decade it seemed that evidence from quasars supported the steady state model and correspondingly toppled the so-called big bang model, along with its insufferable moniker.[11]

In 1965, however, another discovery was made. This too, like quasars, was unexpected at the time – but, as it turned out, it was not unpredicted. At Bell Labs in New Jersey, two scientists were bouncing radio waves off a satellite, trying to fine-tune the signal with an antenna shaped like a large horn. Robert W. Wilson and Arno Penzias could not get a clear signal,[12] so after eliminating what they thought were all the sources of interference, they essentially tuned-in on just the remaining background noise, as a means of locating its source. Not only was the strange noise still there, but they found it came equally from all directions in the sky, day and night. At one point they climbed into the horn and found a nest of pigeons; they shooed away the pigeons and cleaned-up the mess, believing they found the source of the noise. But the enveloping noise persisted. It seemed to come from everywhere.

They mentioned this puzzle to colleagues within the physics grapevine, and the answer was found about forty miles away at Princeton University. While Penzias and Wilson were cleaning pigeon droppings, three astrophysicists, Robert Dicke,

---

[11] Topper [198], pp. 77–80.

[12] Wilson, an American, was born in 1936. Penzias was born in Germany in 1933.

Jim Peebles, and David Wilkinson[13] calculated that if the universe began with a big bang, then some background radiation should remain in the microwave range,[14] which was what Penzias and Wilson measured with the horn antenna at Bell Labs. This was the answer to their puzzle, and Penzias and Wilson later got a Nobel Prize for their accidental discovery. Unlike quasars, whose discovery led to a search for their cause, the discovery of the microwave background radiation had a hypothetical cause waiting to be merged with the discovery.

Interestingly, unbeknownst to all of them at the time, much earlier, about the time that Bondi, Gold, and Hoyle were speculating on the steady state idea, another trio of scientists was working on the expanding model. George Gamow, Ralph Alpher, and Robert Herman[15] had also calculated a residual microwave energy. So, why didn't anyone look for this background radiation over the previous decade-and-a-half since it was postulated? It may be that although such speculation around the expanding model was occasionally pursued, the model was not taken seriously enough for anyone to make the experimental effort to confirm it.

This measurement of what became known as the cosmic microwave background radiation – or other shorter versions and variations of the phrase – was one of the major discoveries of the 1960s. Yet, contrary to some textbook accounts of the history, it was not immediately seen as proof of the expanding model. The steady state model was well-entrenched in many astronomy departments, and into the 1970s the redshift of galaxies was still explained among many astronomers by other causes than a Doppler shift.[16] Before continuing that story, however, there remains the third key episode from the previous decade: the postulation of black holes in the late-1960s.

Although the idea of a concentration of matter so strong that nothing can escape its gravity goes back to the late-eighteenth century – and especially since 1915 when Einstein presented his general relativity, when further mathematical calculations were made regarding various possible radii of such compressed masses – the idea was not taken serious until the discovery of neutron stars (or pulsars) in 1967,[17] the same year physicist John Wheeler coined the term black hole.[18] As mention before,

---

[13] Dicke, American (1916–1997). Peebles, Canadian–American, born 1935. Wilkinson, American (1935–2002).

[14] If the universe did begin with this "bang," it made no sound, not only because there was no sound (there is no sound in a vacuum) but it was not an explosion; rather it was a concentration of repulsive energy in one "place" that immediately expanded. This also means that the often used metaphor of calling the remaining radiation the "echo" of the big bang is erroneous.

[15] George Gamow (1904–1968) was mentioned above in a previous footnote on Einstein's biggest blunder (Chapter 22, #1). Ralph A. Alpher (1921–2007), also mentioned in that footnote, was a Russian-Jewish immigrant who was a doctorial student of Gamow in the late 1940s, and went on to work with him on expanding model. Robert Herman (1914–1997), American physicist, Ph.D. from Princeton University, had an eclectic career in theoretical and applied physics, working as well in engineering. He turned to making sculptures in his latter years.

[16] Kragh [121], Chap. 7; Topper [198], pp. 77–80.

[17] They were discovered by the British astronomers, Jocelyn Bell Burnell (born 1943), who was a postgraduate student at the time, and her thesis supervisor Antony Hewish (born 1924).

[18] American physicist (1911–2008).

from about the 1930s, when the life cycle of stars was beginning to be understood as a process of nuclear reactions, it was realized that different stars have different cycles, depending primarily on their mass. Our Sun, an average star, is about halfway thought its ten- to twelve-billion year cycle, and will end by growing out to the orbit of Jupiter and then collapsing into an inert white dwarf. A much larger star (about one-and-a-half times our Sun) will end as a neutron star, and send out pulses of energy as it rotates; hence the name pulsar. An even larger supersized star (over three times the mass of our Sun) will swell into a red giant and end with a massive explosion, called a supernova, and then collapse and contract into a black hole.[19] (Another important by-product of this event is the creation of the higher elements of the periodic table, an idea postulated much earlier by Hoyle but not taken seriously at the time.) It was in the late 1960s that the important distinction between novas (stars that occasionally pulse in brightness) and exploding supernovas was understood. The combination of the confirmation of pulsars and the adoption of the clever (if also misleading) term, black hole, brought the notion of the existence of an enormous concentration of grav-ity into a very small space into the forefront of astronomy.[20]

Serious searches for black holes commenced in the last quarter of the century, and by the turn of the millennium there was strong evidence of their reality. Although, by definition, black holes cannot be seen directly, their existence can be detected when they are near a large star and they siphon-off the star's energy: a stream of light emanates from the star, rather like a cone, which diminishes in size toward a point, and then disappears into...well, a black hole.

The decade of the 1960s saw the introduction of quasars, black holes, and the cosmic background radiation into the mix of astronomical things, within the frame-work of the steady state model. The 1970s, consequently, was a transition period with the gradual elimination of the steady state model, except for some diehard believers. In 1980, for example, astronomer Vera Rubin[21] was asked to write a sum-mary article for the journal *Science*. While making passing reference to the steady state model, she wrote: "Most astronomers accept as a model a universe which has expanded and cooled from an initial hot, dense state." In a later issue of the journal a letter supporting the steady state model was published accompanied by a non-Doppler explanation of redshift; in addition, the author cast doubt on Rubin's asser-tion that "most astronomers" believed the expanding model.[22] Even so, by the mid-to late-1980s there were few established astronomers still clinging to the steady state

---

[19]Eisenstaedt [60], p. 313. Eisenstaedt's book is an erratic yet brilliant and insightful history of black holes and general relativity. For a short overview, which also reveals the role of Robert Oppenheimer that Eisenstaedt doesn't even mention, see Bernstein's essay, "A Brief History of Black Holes," in Bernstein [9].

[20]The term black hole is really a metaphor, but unfortunately it is often taken literally, especially within science fiction. As with the term light-year, black hole has become a metaphor ubiquitous in popular media.

[21]American, born in 1928, she was the first woman permitted to use the giant telescope on Mt. Palomar.

[22]Rubin [176], pp. 63–71.

Fig. 24.1 The great wall of galaxies. The distribution of the galaxies was discovered to be clumped together in places, rather than spread evenly over space. The diagram depicts two cones in the northern and southern hemispheres. The gap between them is due to the Milky Way, plus gases and dust, obscuring our view

idea. The microwave background was by then interpreted as, at once, supporting the expanding model and falsifying the steady state.[23]

About the time that this consensus among astronomers was materializing around the expansion model, an important celestial survey of all visible galaxies in both hemispheres was coming to fruition. Mapping hundreds of thousands of galaxies as points within two cones (pointing north and south, with the Milky Way in-between blocking out the view of other galaxies), the resulting picture of the heavens exposed a puzzle (Fig. 24.1). The galaxies were not distributed as evenly throughout space as expected; instead they clumped and grouped together, especially along two distinct paths, one in each hemisphere. The paths were metaphorically called the great walls of galaxies, and were a source of momentary doubt for the expansion model – a model that implied a more even distribution of matter because the background radiation came uniformly from all directions.

Or did it? Remaining committed to the expansion model, astronomers proposed a further look at this radiation – to fine-tune the noise by listening to it from a satellite in orbit, far from earthly interference and above the atmosphere. Named COBE (an acronym for COsmic Background Explorer) the satellite was launched in November, 1989 and measured radiation from the entire three-dimensional sphere of the heavens. The computer generated image (Fig. 24.2) [24] used different colors for different temperatures, and clearly displayed fine ripples in the radiation; that is,

---

[23]Topper [198], pp. 80–81. See also the chart in Kragh [121], p. 379, comparing polls conducted among American astronomers in 1959 and 1980, showing the dramatic change.

[24]The result is an elongated oval; think of a two-dimensional map of the spherical Earth.

Fig. 24.2 COBE. An image of the background radiation of the universe from the Cosmic Background Explorer satellite. This is a 2-D projection of the "sphere" of space revealing different temperatures or ripples in the radiation, which are represented by different shades of grey in the computer generated image

rather than showing one uniform color, the image revealed blotches of different colors across the oval field (which are different shades of grey in my figure).[25] This is what the astronomers expected to see if not long after the initial big bang, with matter spreading-out while space and time were forming, clumps were already beginning to form as gravity was pulling together matter into stars, later galaxies, and even groups of galaxies – except not necessarily uniformly. The so-called great walls of galaxies were no longer a problem. COBE secured the expansion model for the start of the millennium, and along with the triumph of that model came the common adoption of the term big bang.

What of quasar? As seen, the problem of their enormous energy was considered evidence for the steady state model in the 1960s. In recent years a consensus is forming around an explanation that brings into play black holes. According to the latest model, a quasar may be a galaxy with a spinning black hole at the center. As the stars close to the hole are sucked-in, they travel near light-speed and from the massive resulting friction they emit radiation (from infrared to x-rays) as they enter forever the spinning black hole. This would explain the extreme energy, for the hole could ultimately gobble-up the entire galaxy for fuel, and the short period of fluctuation of radiation would be due to its spinning. The model nicely combines two entities from the 1960s.

As Hubble's law was further recalibrated, the age of the universe began to settle around twelve to fourteen billion years on the big bang model. That model, however, was really a generic term engaging the notion of an expansion that commenced after the initial concentration of energy "in the beginning."[26] What, however, about the process of expansion itself? How fast was the expansion? Has it changed? How long

---

[25]The reader may wish to view a colored image on the Internet. An image search for COBE will easily bring one up.

[26]Technically, this means that the spectral shifts in Hubble's Law are due to the expansion of space itself, and therefore properly speaking they are not Doppler shifts, since they are not directly related to speed. Nonetheless, it is common to ignore this technicality and call the redshifts of distant galaxies Doppler shifts.

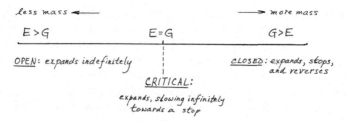

Fig. 24.3 Variations of the expanding universe. Different ratios of the amount of matter (gravity) verses the force of expansion (energy) results in three possible models

can it go on? What was there before the initial concentration of energy? In attempting to answer some of these questions, there arose several variations of the expansion model.

Return to the balloon analog. In blowing-up a balloon, two forces control the process: the force of the air from our lungs, and the resistance due to elasticity from the balloon's material. In the universe the force of the air corresponds to the initial energy (**E**) of the big bang producing a force of expansion, and the opposing elasticity corresponds to the attractive force of gravity (**G**) resisting this outward force. As the universe expands, it cools; this leads to energy transforming into mass (according to $E = mc^2$)[27] resulting in more matter continually being formed. More matter means more gravitational attraction, which is really due to the curvature of space from general relativity. The different ratios of these two powers (**E** to **G**) correlate to different variations of the expansion model. To see this: consider the horizontal line (in Fig. 24.3) as a continuum of the model's variations, where to the left there is less mass in the universe and to the right more mass. To the left, **E > G** (i.e., the outward energy is greater than gravity) and to the right **G > E** (gravity's attraction is greater than the energy). To the left, therefore, the universe will continue to expand indefinitely; this is called an open universe. To the right the universe will initially expand outward but eventually come to a stop; gravity then takes over and the process is reversed, so that the universe closes back in on itself. This is called a closed model.[28] Finally, what about the middle case, where the outward and inward forces balance, such that **E = G**? In this case the expansion proceeds as in the open model but slows down toward a stop, taking an infinite time. This is called a critical model and is the one initially preferred by most cosmologists based on what was known about the early universe.

Which of the three variations of expansion is our universe? It mainly depends upon the amount of mass (matter) in the universe.[29] Initial calculations used only the

---

[27] The process is a phase transition, like water turning into ice.

[28] There are also variations of this specific variation in that it is not known what happens after the first cycle: does it end there, or does it start over again in an endless series of cycles, and if so are these cycles identical or not. Due to the probability of minor fluctuations over every cycle, the consensus is that they would not be identical.

[29] More correctly, it is the density of the universe.

light-spectrum, and an open universe was the result. However, the idea of there being more matter in the universe than what we know from the light-spectrum alone was put forward in the 1933 by Zwicky.[30] Consider the Andromeda galaxy: its mass can be measured by the amount of matter we see from the spectrum of light. Another way of measuring its mass is from it rotational motion. When it spins there are centrifugal forces outward which are balanced inward by the pull of gravity. By measuring the rotation and distribution of matter, it is therefore possible to measure the amount of mass that holds the galaxy together. Using this other method to measure Andromeda's mass Zwicky found its rotational mass was ten times its visible mass. Since he believed that the rotational mass was real mass, he declared that 90% of the galaxy's mass was made up of invisible matter, what he called "dark matter."[31] Zwicky went on to discover this discrepancy in other galaxies, but nothing much came of this work, since it did not fit into any theoretical framework at the time. His discovery was filed as an anomaly, so to speak.[32] In the late-twentieth century, however, Zwicky's discovery was revived as a means of bolstering the critical model; accordingly, moving to the universe's mass closer towards the critical mass in Fig. 24.3.

Nonetheless, just as astronomers and physicists speculated about possible sources of this dark matter, supernovae at the far edge of the universe coughed-up a clue to another problem. There are four types of supernovae, known by their different spectra, and caused by different processes of formation. In the case of the brightest supernova type, when that star reaches its maximum brightness, the luminosity is the same magnitude for all supernovae of this type, as far as astronomers know. If true, this is extremely important because the luminosity is then an absolute number and by using the inverse-square law of diminishing light intensity, its distance can be calculated. Such cancelations were and are being made using this important type of supernovae, which are found in the most distant galaxies. Surprisingly, the calculations showed that the universe is expanding at an accelerative rate. Instead of them slowing down toward a critical universe, as hoped, they appeared to be receding faster and faster. The result is what has been called an inflationary universe.

If this acceleration of the universe is actually happening, something else is at work propelling this expansion, such as a previously unknown repulsive force within the universe. Such a force must be part of space itself: as the universe expanded from its start as one clump of energy (the big bang), and mass (from $E=mc^2$) spread out, more space emerged as the universe expanded. If space contains the repulsive force, then it comes into effect at a considerable time into the universe's history when there is a greater spatial force than a material force. Not irrelevant in all this is that space in Einstein's general relativity was not just a passive receptacle for matter but was actively involved in curving or warping into gravity. Contemporary cosmologists have named this property of space, which may be the source of expansion, dark energy – another concept added to that of dark matter.

---

[30] He, recall, also proposed the gravitational drag idea about light to explain redshift without Doppler's principle.

[31] This is an equivalent English for his original German, "dass dunkle Materie," Zwicky[220], p. 125.

[32] For an excellent history (and pre-history) of dark matter, see Trimble [201].

Then in June, 2001, another satellite, WMAP (Wilkinson Microwave Anisotropy Probe)[33] was sent into orbit to measure the microwave background even more accurately than COBE. It has set the age of the universe at 13.73 billion years, and further determined that dark matter makes up 23.3% of the matter in the universe, and dark energy 72.1%, the latter being the cause of the speeding-up of the expansion rate. The remaining 5.1% is all that is left of the ordinary matter we see around us everyday and that is, in fact, us.[34] Presently, a key task for observational astronomers is to find the sources of dark matter and dark energy. In the meantime, we are still in the dark.[35]

Which, you may have guessed, brings us back to Einstein: for scientists have obviously noticed that there is an interesting similarity between the introduction of a repulsive force (dark energy) into this recent expanding universe and Einstein's cosmological constant. Indeed, some cosmologists relate the dark energy to Einstein's original idea, which, they now say, he erroneously abandoned. If this model turns out to be correct, Einstein's greatest blunder was no bungle after all.

* * *

As seen, in the early 1950s, a few years before he died, Einstein called the cosmological constant the biggest blunder of his life. Interestingly, or ironically, Lemaître had continued using the cosmological constant as a sort of vacuum energy in space.[36] It appeared, for example, in his paper contributed to the 1949 Schilpp two-volume collection of essays devoted to Einstein's impact on the twentieth century.[37] At the end of Volume Two Einstein replied to selected essays. On Lemaître's essay he dismissed the argument as not "sufficiently convincing in view of the present state of our knowledge" and rejected the cosmological constant as "unjustified" theoretically.[38] Further, Farrell quotes from a letter of Einstein to Lemaître on the essay:

> Since I have introduced the [cosmological] term I had always a bad conscience. But at the time I could see no other possibility to deal with the fact of the existence of a finite mean density of matter. I found it very ugly indeed that the field law of gravitation should be composed of two logically independent terms which are connected by addition. About the justification of such feelings concerning logical simplicity it is difficult to argue. I cannot help to feel it strongly and I am unable to believe such an ugly thing should be realized in nature [emphasis added].[39]

---

[33] The satellite was named in honor of David T. Wilkinson, mentioned above along with Dicke and Peebles, after his early death due to cancer. He made fundamental contributions to many major experiments on the microwave background radiation, including COBE and the WMAP.

[34] http://map.gsfc.nasa.gov/

[35] Despite the consensus on interpreting dark matter, dark energy, and the inflation of the universe from the big bang model, there are some challenges. For example, Lerner and Almeida [131] argue that these are "hypothetical entitles" that support the big bang model, without an open debate for alternative models.

[36] Farrell [61], pp. 116–117: McVittie [140], p. 296.

[37] Lemaître [130].

[38] Einstein in Schilpp (ed.) [179], Volume Two, pp. 684–685.

[39] Farrell [61], quoted on p. 169. It is not clear to me if Farrell or Einstein added the emphasis. The letter is dated September 26, 1947, and is listed in the Einstein Archives in Jerusalem, but unfortunately it is not yet available on-line.

This probably was Einstein's last written word on the topic, before he reiterated the point in his "blunder" statement several years later. Note that he emphasized the aesthetic predicament associated with the constant, and admitted that ultimately such matters are of a subjective nature – that is, strongly felt but difficult to justify. Einstein was clearly aware of his inner world's role in judging scientific ideas.

At this point in the book's narration – which, indeed, has strayed far beyond the death of Einstein, and yet furtively looped back to him and his cosmological constant – it would be prudent to leave behind cosmology as a topic, a science still in flux and to be continued…., while we, instead, return to where we left Einstein in the 1920s, striving to extend his general theory to include electricity – and even more – in his quest for unification.

Part V
Exodus: Quest for a Unified Field Theory

> *Of course it would be a great advance if we could succeed in comprehending the gravitational field and the electromagnetic field together as one unified conformation. Then…the whole of physics would become a complete system of thought….*
>
> (Einstein's Leiden lecture of 1920).[1]

During the third Caltech visit, in January of 1933, Adolf Hitler and the Nazi party were elected in Germany and thus began a brutal dictatorship that lasted a dozen years. The previous fall, as Albert and Elsa were leaving their country cottage and returning to Berlin, he told Elsa, "Take a good look at it." She asked why. "You will never see it again," he is reported to have said.[2] They never returned. Indeed, after departing for California, they never set foot on German soil again.

The Einsteins ultimately settled in the United States. There were two major choices: on the west coast (with Millikan, as seen, hoping Einstein would remain at Caltech) and on the east coast, where the new Institute for Advanced Study was recently created in Princeton, New Jersey, nearby but autonomous from Princeton University.[3] Einstein accepted the Institute's offer, which was a research position with no teaching duties. He remained there for the rest of his life, pursuing mainly a twenty-two year quest – the main topic of this Chapter.

My choice of *Genesis* for the title for Part I obviously engages an intentional biblical allusion, both in terms of the concept of origins and beginnings, and as well as allusions to the role of light in Einstein's thought experiment and light on the first day of creation – "and there was light…, the light was good" – in the book of *Genesis*.

---

[1] Einstein [53], pp. 22–23. The same translation is reprinted in *Einstein Papers*, Vol. 7, Doc. 38. An alternative translation is in Einstein [42], pp. 110–111. "It would, of course, be a great step forward if we succeeded in combining the gravitational field and the electro-magnetic field into a single structure…. [Then] the whole of physics would become a completely enclosed intellectual system….".

[2] The incident is told by Frank [67], p. 226.

[3] It is often erroneously assumed that the Institute is part of Princeton University because initially it was housed in the University's mathematics department until the Institute buildings were completed in 1939. They are less than two miles apart.

The choice of *Exodus* as the title for this last Part engages multiple meanings too. With reference to the topic of this Part: to be seen, in his work at the Institute there was the rejection of the prevalent interpretation of quantum physics (entailing his exodus from that branch of physics), and at once his intellectual voyage toward a unified field theory fusing all of physics.

The most direct meaning clearly pertains to Albert and Elsa exiting Germany under threat of persecution, and moving to America, the Promised Land.

For this they were part of what became a mass exodus from Germany, what I believe was one of the most massive "brain drains" in modern history. By the spring of 1933 numerous professors were dismissed, mainly because of their Jewish ancestry or anti-Nazi activities. The numbers are staggering: 10% of all professors lost their jobs in Germany for having "Jewish blood"; in mathematics it was 20%, and in physics 26%. The famous Hungarian-American mathematician John von Neumann was in Germany in the summer of 1933 and wrote of the "horrible situation" in the universities, calling it "German madness," and predicting that it "will ruin German science for a generation – at least...."[4] He was right: by the spring of 1936 more than 1,600 scholars (one-third of them scientists) were gone from German institutions, the majority going to the United States. The aftermath of this brain drain may be measured by comparing the number of Nobel prizes in science between the two countries before and after the Second World War. From the inception of the Prize in 1901 to the War, thirty-five Germans were recipients compared to fifteen Americans. After the War through 1959, only eight Germans got the Prize compared to forty-two to U.S. scientists, many of them surely being refugees from Germany. Madness, indeed.

There is, as well, an important story from Einstein's life that needs to be fleshed-out involving his rescue of refugees (mainly Jews and other dissenters) from Nazi Germany in the 1930s, when it was difficult for Germans to enter the United States due to restrictive immigration rules. Einstein wrote affidavits on behalf of perhaps 100 individuals and paid for their rescue, often by providing a financial guarantee. Leopold Infeld, in Poland at the time, and who received a fellowship to the Institute for Advanced Study in 1936 based on Einstein's support,[5] later wrote: "Everyone had a testimonial from Einstein."[6] In a letter to his sister in 1938 Einstein wrote that

---

[4] Quoted in Topper [197], p.141.

[5] When the fellowship ran out Einstein offered to support Infeld but he refused. Instead, Infeld suggested they collaborate on a book. Thus was born *The Evolution of Physics*, 1938, which became a best seller, with the royalties supporting Infeld until he found a position at the University of Toronto in 1939. Infeld [106], pp. 307–322.

[6] Quoted in Neffe [149], p. 377.

"Miss Dukas and I run a kind of immigration office." Neffe says that Einstein "paid a small fortune" toward this effort.[7]

But our focus here is science, not politics[8] – specifically Einstein's intellectual effort towards the Promised Land of a final unification of physics.

---

[7] Quoted in Neffe [149], p. 377. Also Frank [67], pp. 275–278; and Sayen [178], Chap. 6, esp. pp. 112–116.

[8] Nonetheless, Einstein got heavily involved in political matters, from which he garnered one of life's lessons: "People flatter me a long as I do not get in their way. But if I direct my efforts towards objects which do not suit them, they immediately turn to abuse and calumny in defense of their interests. And the onlookers mostly keep out of the light, the cowards!" Quoted in Einstein [44], p. 69. This was probably written in the early 1930s. For a case study of politics and physics in Einstein's friendship with Friedrich Adler, see Galison [77].

Chapter 25
Roots of, and Routes Toward, Unification

In seeking a unity of forces in nature Einstein was drawing on a tradition going back at least into the previous century. As seen in Chap. 4, nineteenth century physics was awash in ideas of conservation, transformation, and unification – all three coupled into a conceptual whole. Regarding the specific forces in Einstein's quest, the framework goes back to Newton's trilogy of space, force, and matter. Kant's subsequent unification was based on his reduction of matter to force, reducing the trilogy to a duality of force and space. Kant's concept of force then morphed into energy, and Einstein's $E = mc^2$ changed the duality to mass-energy and space. When gravity became warped space (really space-time), gravity was accounted for. A beautiful unification.

* * *

In the meantime, a parallel conceptualization was evolving around electricity and magnetism. First, there was the formal analogy between the inverse-square law of gravity and the inverse-square laws of electric charges and magnetic poles – namely, Cavendish's and Coulomb's experiments, respectively. Second, these laws were associated with occult powers. Newton had justified his action-at-a-distance explanation of gravity by an analogy with the obvious action-at-a-distance fact of nature for electricity and magnetism. Yet it was through the work of Faraday on these very forces that the field model arose, whereby spooky forces acting instantaneously between electric charges or magnetic poles across space were replaced by force fields acting within space and over time. Maxwell's equations were the crowning mathematical expression of this model, which after his death revealed that light too is embraced by this model – namely, that all light, visible and not, is merely electromagnetic radiation. Maxwell was reluctant to accept the autonomy of the field independently of its grounding in a material base such as an aether – Faraday's conviction for the reality of the field, not withstanding – but later scientists, Einstein for one, built further on the self-sufficiency of the field in their interpretations of Maxwell's equations. Since the electromagnetic field was shown to contain energy, Ostwald, as seen, went so far as to reduce everything, in the spirit of Kant, to energy alone.

D.R. Topper, *How Einstein Created Relativity out of Physics and Astronomy*, Astrophysics and Space Science Library 394, DOI 10.1007/978-1-4614-4782-5_25,
© Springer Science+Business Media New York 2013

In the end, Einstein's ideas surely were part of the history of these conceptualizations when he reduced gravity to the geometry of space in the theory of general relativity. As just noted, Newton had justified gravitational action-at-a-distance by an analogy with electricity and magnetism. So Einstein, by sort of turning Newton's analogy around, employed Faraday's field model from electricity and magnetism, and applied it to gravity, replacing Newton's spooky action-at-a-distance with a field theory of gravity. Late in life Einstein spoke of this "emancipation of the field concept" from the assumption of an aether as "among the psychologically most interesting events in the development of physical thought" – a very strong statement, with which I nonetheless concur.[1] In his prime Einstein used the field in a fashion similar to Ostwald, although he never embraced energeticism. Here is an outline of his argument as put forward in the book he wrote with Infeld.[2]

The introduction of the field concept in the nineteenth century imparted a matter/field dualism to physics. "But the division into matter and field is, after the recognition of the equivalence of mass and energy, something artificial and not clearly defined." From this followed a rhetorical question: "Could we not reject the concept of matter and build a pure field physics?" Sounding exceedingly similar to Kant filtered through Ostwald, Einstein noted that matter, as it "impresses our senses," is "really a great concentration of energy into a comparatively small space." This meant that we may regard matter as "regions in space where the field is extremely strong." The result was a theory that explained "all events in nature by structure laws valid always and everywhere," and which eschewed the dualism of matter and field, such that the field is "the only reality."[3] From this conceptualization it was not much of a leap to endeavor to unite gravity and electricity within the framework of "a pure field physics," which, recall, was Faraday's final quest. This conceptual framework was another source of the unification notion.

* * *

It is true that relativity for the non-technical reader is usually presented as a theory of mechanics – conceptualized in terms of inertial systems, the speed of light, the reinterpretation of time, mass, space, and so forth, culminating in the finale of a new cosmology. Einstein himself initiated this approach to the subject: probably only close readers of his popular account of 1917 were aware of how

---

[1] Einstein [49] [1952], Appendix V, p. 146.
[2] Einstein and Infeld [59] [1938]. In the introduction to the 1961 edition of the book, Infeld called Einstein the "chief author" of the book. For the origin of the book and the collaboration, see Infeld [106], pp. 308–321, where he implies that it was a bit more collaborative than implied in the phrase "chief author." I will, nevertheless, refer to the ideas in the book as Einstein's, since Infeld does affirm: "We always reached some kind of compromise," p. 315.
[3] Einstein and Infeld [59] [1938], pp. 242–243.

steeped the theory was in electrodynamics, since he diminished the significance of electrodynamics by not even mentioning the Faraday experiment as a source of the relativity principle.[4] As well, Einstein pursued this purely mechanical approach in his technical work on the theory; for example, in 1934 he delivered the prestigious Josiah Willard Gibbs Lecture of the American Mathematical Society in Pittsburgh, where he presented a derivation of $E = mc^2$ based solely on an elastic collision between two identical masses. It is clear from the details of the lecture that the motivation was to derive relativistic concepts such as energy and momentum from purely mechanical principles, free of electromagnetic theory.[5] Nonetheless, despite these various expositions of relativity that disengaged mechanics as an integral foundation of the theory, the historical evidence is that theory was immersed in electromagnetism from the start. As noted before, those who are introduced to relativity only from popular accounts are often exceedingly surprised to find that the actual title of the 1905 paper was, "On the Electrodynamics of Moving Bodies," not something like, "On the Theory of Special Relativity."

This brings us back to the fact, as emphasized in Part **II**, that Faraday's ideas played a critical role in the genesis of relativity. Firstly, Faraday's experiment with the magnet and wire was the only experiment that Einstein explicitly referred to in order to support the relativity postulate. The 1905 paper itself likewise showed that Maxwell's equations are covariant for all inertial systems. Furthermore, at the end of the 1907 review article, Einstein put forth the possibility that light (namely, electromagnetic radiation) was bent by gravity; here was a direct connection between the two, a topic that was pursued through the general theory of 1915, when he got the correct value for a star's visual shift by the Sun during an eclipse. Finally, Maxwell's equations appear again in the magisterial 1916 review article on general relativity.

* * *

Thus, from the above roots of, and routes toward, unification, we find that following the 1917 cosmological application, Einstein began the task of unifying gravity and electromagnetism. That these byways led to the quest was clearly expressed in the quotation from his 1920 Leiden lecture, which is used as the epigraph introducing this chapter. "Of course it would be a great advance if we could succeed in comprehending the gravitational field and the electromagnetic field together as one unified conformation. Then…the whole of physics would become a complete system of thought…."

Prior to commencing a discussion of Einstein's quest, there is one more important point to make clear. Even though electromagnetism played a key role throughout the development of special and general relativity (as reiterated again),

---

[4] Einstein [49] [1917], passim.

[5] The meeting was held at Carnegie Institute of Technology (now Carnegie-Mellon University). The presentation was probably his first public appearance in his newly adopted country. See, Topper and Vincent [200] where we put forth a reconstruction of the lecture.

the role did not constitute an integration of the two. Otherwise, without doubt, the quest would be over! Stated explicitly, the force of gravity and the electric and magnetic forces were still not integrated into one theoretical structure. It is true that Maxwell's equations for empty space did not contradict relativity, but it is a huge step (indeed, one still not fathomed by any physicist even today) to put gravity and electromagnetism together under one all-encompassing theory. Let's see how Einstein tried…and tried…and tried….

Chapter 26
1931: Caltech, Once More

During his first Caltech visit, in the lecture delivered to the observational astrono-mers (Photo 18.1; Fig. 18.1)[1] who both confirmed his relativity theory and changed his mind about the static universe, the actual topic of this lecture was neither of these matters: rather it was his then present obsession with the unified field idea. So we return, once more, to the first Caltech visit, when Einstein not only listened to the astronomers but he also told them a thing or two – or, at least, he tried to.

As seen, what the experimentalists told him not only supported his relativity theory – much to his pleasure, I'm sure – but also led to his rejection of two closely held concepts, the cosmological constant and Mach's principle. The discarding of these from his theory was a result of his acceptance of a Doppler interpretation of Hubble's work, and his abandonment of the static model. Yet, independently of the cosmological problem of a finite or an infinite, a static or expanding universe, I think there was another path to the rejection of Mach, this one coming from his idea of unification – and I contend that it imparted the final death knell of the principle. Here is my argument.

A key conceptual feature (if not <u>the</u> conceptual feature) of general relativity was the explanation of gravity as a curvature of space (or space-time). Warped space eliminated the need for action-at-a-distance, which was a concept of forces that har-kened back to occult powers, and which Einstein himself later admitted was "spooky." But this occultism of gravity was an essential part of what Einstein called Mach's principle, with inertia being caused by the attraction (at immense distances, no less) of the fixed stars. It seems that in Einstein's mind this led to a trade-off: on the one hand, Mach's principle embraced the relativity of all motion and as a bonus explained inertia, although it also carried the burden of spooky action-at-a-distance, which con-tradicted his commitment to a field explanation; on the other hand, without the prin-ciple, we are back to the mystery of inertia. Here was how he put it the Leiden lecture of 1920: "It is true that Mach tried to avoid having to accept as real something which is not observable," that is, absolute space; he did this by substituting into "mechanics

---

[1]Except for Michelson.

D.R. Topper, *How Einstein Created Relativity out of Physics and Astronomy*, Astrophysics   199
and Space Science Library 394, DOI 10.1007/978-1-4614-4782-5_26,
© Springer Science+Business Media New York 2013

a mean acceleration with reference to the totality of the masses in the universe in place of an acceleration with referent to absolute space. But inertial resistance [which is] opposed to relative acceleration of distant masses presupposes action at a distance." True enough: and there it is, in that last phrase, "...presupposes action at a distance." In short, he appears to be brooding over this important question: Is Mach's explanation of inertia worth the retention of occult forces? Einstein's answer was written in the third person, but my guess is that he was speaking of himself when he wrote that "the modern physicist does not believe that he may accept this action at a distance."[2] He then went on to speak of a reintroduction of a special form of aether to account for inertia, but I find the terminology essentially a semantic ploy; I interpret this surrogate aether in light of his realization that space as expressed in general relativity is a physical entity, not merely – so to speak – nothing. He stated it this way a few pages later: "According to the general theory of relativity space is endowed with physical qualities; in this sense, therefore, there exists an aether."[3] It was unnecessary and unfortunate for him to bring back the term aether at this time, since he so summarily dismissed it in the 1905 special relativity papers. Indeed, he concluded the Leiden lecture by emphasizing that this aether did not consist of any material substance, nor did it have the property of motion. Undoubtedly, instead, to have just said that space is real would have been enough.[4]

The context of the lecture throws some light on Einstein's rhetoric. He was asked to deliver this formal lecture to the University of Leiden by Lorentz, who continued to embrace the concept of the aether, and Einstein had great respect and admiration for him. Indeed, Lorentz requested that the topic be the aether, so it seems Einstein had no choice.[5]

To be sure, the topic was on his mind at the time. In the important document extensively quoted (twice) in Chap. 12, also written in 1920, where Einstein recounted his thought experiment involving free fall and deduced what became the equivalence principle, he mentioned the aether near the end in the following way. Since general relativity has given physical properties to otherwise empty space, it may be said that the aether has been "resurrected," although in a different "sublime form." Moreover, since this medium has neither substance nor motion, then "the concepts of 'space' and 'aether' flow into each other."[6] So he ended with a metaphor.

---

[2]Einstein [53] [1920], p. 17. The entire essay, "Ether and the Theory of Relativity," is reprinted on pp. 1–24. As pointed-out in the footnote to the epigraph to this chapter, an alternative translation of the Leiden lecture is in Einstein [42]; it is titled "Relativity and the Ether," pp. 98–111. The above quoted passage is essentially the same translation (p. 107).

[3]Einstein [53] [1920], p. 23; Einstein [42] [1920], p. 111.

[4]On this I depart from the opinion of some historians who contend that Einstein seriously brought back the aether in the 1920s. I should also point out that the title of the lecture, "Aether and the Theory of Relativity," or, "Relativity and the Aether," neither supports nor negates their interpretation. An atheist may title her lecture, "God and Science."

[5]*Einstein Papers*, Vol. 7, Doc. 38, footnote 1.

[6]*Einstein Papers*, Vol. 7, Doc. 31; "...*fliessen die Begriffe 'Raum' und 'Aether' zusammen.*"

Furthermore, it looks as if he could not let go of this conceptual problem, for in the Einstein archives there are four manuscripts in 1930 of papers on variations of the theme involving space, field, and aether.[7] Clearly he was still pondering this topic a decade after the Leiden lecture. Two of the essays are available in English translation. The most accessible is "The Problem of Space, Aether, and the Field in Physics," in the continually reprinted book, Ideas and Opinions.[8] In this essay, while discussing the aether in the nineteenth century, he asserted that "physical space and the aether are only different terms for the same thing; fields are physical states of space. For if no particular state of motion can be ascribed to the aether, there does not seem to be any ground for introducing it as an entity of a special sort alongside of space." He went on to mention "the genius of Riemann" [one of the founders of non-Euclidean geometry, mentioned before] who proposed "a new conception of space, in which space was deprived of its rigidity, and the possibility of it partaking in physical events was recognized."[9] As seen, Einstein applied this geometry to derive gravity in his general relativity. That monumental fulfillment, however, was not an end in itself – rather, it initiated the next task. As he wrote:

> Gravitation had indeed been deduced from the structure of space, but besides the gravitational field there is also the electromagnetic field. This had, to begin with, to be introduced into the theory as an entity independent of gravitation. Terms which took account of the existence of the electromagnetic field had to be added to the fundamental field equations. But this idea that there exist two structures of space independent of each other, the metric-gravitational [i.e., the gravitational term]and the electromagnetic [term], was intolerable to the theoretical spirit. We are prompted to the belief that both sorts of field must correspond to a unified structure of space.[10]

This conceptualization was the basis of the unified field theory, with both gravity and electricity emerging out of the same space. The "theoretical spirit" of which he spoke was, certainly, Einstein's alone, and especially noteworthy was his use of the

---

[7]See also Schilpp (ed.) [179], vol. II, p. 721.

[8]Einstein [47] [1930], pp. 276–285. The essay is erroneously dated 1934, since that is the date of the reprint in the Seelig book [47], 1954. A slightly longer version, with some minor variations in the translation, is in the less accessible book, *Essays in Science*, Einstein [42] [1930], pp. 61–77. This latter essay includes seven paragraphs at the end, which were cut in the 1954 book.

[9]Einstein [47] [1930], p. 281; Einstein [42] [1930], p. 68, is essentially the same translation. Bernhard Riemann (1826–1866), German mathematician, made major contributions to both tensor calculus and non-Euclidean geometry; in fact, his curvature tensor was used in Einstein's equation for the general theory. Einstein further implied that Riemann conceived of this distortion of space as having possible physical consequences and hence anticipated Einstein's own theory. Indeed, in the mid-nineteenth century Riemann speculated about a possible unification of forces based on his mathematical work, but it should be emphasized that the physical basis of this idea was the all-encompassing aether. Whittaker [209] [1951], pp. 240–241.

[10]Einstein [47] [1930], p. 285; Einstein [42] [1930], pp. 73–74, is essentially the same translation. I added the emphasis.

term "intolerable,"[11] which clearly betrayed an esthetic foundation to his quest. To him there was only one space, so somehow the discrete fields of gravity and electricity must arise out of different geometries of this same space – for multiple spaces would be an intolerable idea.[12]

There is no evidence that this essay was ever presented publically or published, but another essay with a similar title, written that year, "Space, Aether and Field in Physics," was presented at a conference and published in a journal.[13] As before, he acknowledged the certainty of the aether among nineteenth century physicists because of the interference of light, but noted as well its "strangeness" or "ghostliness." At the same time, the field idea arose as a fundamental reality in electromagnetic theory (recall Faraday's conceptualization), and since "the fields are states of space" then space, field, and aether were possibly all the same thing. Later, from general relativity, came the realization that space had a "structure" and was "changeable." He went on, again, to argue from this that the fields of gravity and electromagnetism may be unified as different structures of space. The essay ended, as in the above 1920 document, with a metaphor – albeit a different and more dramatic one. The earlier metaphor pictured aether and space flowing into one another. Here space "swallowed" ether, so that only space "remains as the sole medium [or support] of reality."[14] This demonstrated, once more, a clear rejection of the aether – precisely as in his 1905 papers that launched his attack on that nineteenth century conception – however much he flirted with it in the 1920s.

For a further buttress to this argument consider the following quotation from an article Einstein published in 1950, where he, once more, raised the question of the need for an aether. Listen to this straight-forward choice: "Since the field exists even in a vacuum, should one conceive of the field as a state of a 'carrier,' [i.e., aether] or should it [i.e., the field itself] be endowed with an independent existence not reducible to anything else?" He reiterated: "In other words, is there an 'aether' which carries the field....?" He went on to mention that in the early years of the field theory, scientists felt a need to base the field on deeper mechanistic foundations (note Maxwell's reluctance to abandon an aether), but today the field stands alone. "Because one cannot dispense with the field concept, it is preferable not to introduce in addition a carrier with hypothetical properties." In short, the field (as Faraday believed) is a real entity unto itself.[15] Bear in mind too Einstein's strong tenet, quoted

---

[11] I searched the on-line Einstein Archives in vain for the manuscript, for I hoped to find the specific German word he used. Recall his use of the term "unbearable" for the asymmetry in Faraday's experiment (Chap. 5).

[12] One wonders what Einstein would make of today's speculation of parallel universes.

[13] Einstein [39].

[14] Einstein [39], pp. 181–184; *verschlingen* (swallow or devour) and *Träger* (medium or support), p. 180.

[15] Einstein [45], p. 259.

above from the 1952 appendix to his popular book on relativity, where the separation of the field and the aether was "among the psychologically most interesting events in the development of physical thought."[16] In that little twenty-two page essay[17] he went on to say that there is no distinction between space and "what fills space"; neither has a separate existence. Take away the field and there is no space, "absolutely nothing [Einstein's emphasis]." Said otherwise, there is "no such thing as empty space."[18] Believing that the only reality is the field kept him going in pursuing the unified field theory, for he ended that essay with the revealing statement that "one [that 'one' being Einstein himself] should not desist from pursuing to the end the path of the relativistic field theory"[19] – which he did, undeniably, to the end.

But I am getting ahead of my story. Here I merely point-out that all of this, I am convinced, harkens back to his ultimate rejection of Mach's principle in the 1920s.

More importantly, Einstein fully realized this problem by the early-1920s. I think it began not long after the completion of the general theory late in 1915, when he commenced his quest to unite gravity and electromagnetism. As seen in Chap. 4 and reiterated above, nineteenth century progress in understanding electricity and magnetism was based heavily on the rise of field theory, which at its core was a rejection of action-at-a-distance. A crowning achievement was Maxwell's equations, four mathematical expressions of the behavior of electricity and magnetism from a field point of view. As well, these equations were even integrated into special relativity; in Part **II** we saw that Maxwell's equations were covariant between all inertial systems. Remember too that Einstein was enamored by the field model, and thus there were conceptual reasons to ditch Mach's principle, tainted with occultism, in the quest for a field theory uniting gravity and electromagnetism. Thus, by the early-1930s, Einstein wrote: "There was [initially] something fascinating about this idea [of Mach] to me, but [in the end] it provided no workable basis for a new theory."[20] This may be what Einstein had in mind when he wrote in his autobiography in the late-1940s that for Mach's idea to be a "truly reasonable theory" it must explain inertia by the interaction of masses according the Newton's mechanics (which I interpret as implying that they must be acting-at-a-distance), and this "does not fit into a consistent field theory...." Without a doubt, it does not; in fact, it contradicts the field model.[21] Henceforth Mach's principle disappeared from Einstein's scientific life, after having had such a grip on him for so many years. As a result, as Pais affirmed, "the origin of inertia is and remains the most obscure subject in the theory of particles and fields."[22]

---

[16]Einstein [49] [1952], Appendix V, p. 146.
[17]Einstein [49] [1952], Appendix V, pp. 135–157.
[18]Einstein [49] [1952], Appendix V, p. 155.
[19]Einstein [49] [1952], Appendix V, p. 157.
[20]Einstein [47] [1933], p. 286.
[21]Einstein [51] [1949], p. 27.
[22]Pais [162], p. 288, emphasis his.

This change in Einstein's opinion of Mach's principle is, however, only half of the story related to the first Caltech visit. The other half is what Einstein said to the astronomers, to which we finally turn.

* * *

This brings us back to the group photograph (Fig. 18.1) that was taken after Einstein's lecture. Alas, what Einstein specifically spoke about is not recorded in any text.[23] All that remains is the equation on the blackboard behind him.

There were three other pictures taken after the lecture. All are close-ups of Einstein alone with his equation, and posed holding a piece of chalk as if he were writing the equation at the time the picture was snapped. A close look at the blackboard reveals that the equation was actually erased after the lecture (there is ghost image of the same equation behind it), and then rewritten by Einstein, probably at the request of the photographer(s?), who wanted an iconic image of the scientist writing an equation.[24]

Like an archaeological shard on a barren site, this equation is the only fragment of the lecture. Although there are several interpretations of what Einstein spoke about at the time, I assert that he presented the topic that obsessed him at the time – namely, the quest for a unified field theory.[25]

So, what is this equation that Einstein wrote on the blackboard, spoke about, erased, and then rewrote again for the benefit of the photographers – and, most importantly now, for posterity? It is $R_{ik} = 0$. That's it, just $R_{ik} = 0$. Well, actually there is more, for a closer look shows that to the right there is a question mark.[26] The complete remnant of the lecture, therefore, is this: $R_{ik} = 0$? An equation followed by a question mark. Not much to go on. Even so, let us probe. First consider the equation without the question mark.

Appearing even simpler than $E = mc^2$, $R_{ik} = 0$ is, in fact, an extremely complex formulation that encompasses not only calculus, but differential equations, and tensor analysis. In contrast, $E = mc^2$ is merely algebra. Put another way, $R_{ik} = 0$ is a compact tensor notation that packs together many larger differential equations into one; for example, for the general theory of gravity alone, the equation can represent up to ten distinct equations. How then can we understand the physics underlying all this – since physics is the topic of this book, not mathematics – and thus penetrate the apparent mathematical barrier? A metaphor from Einstein himself will help.

But first let's begin with general relativity. The essence of the theory of 1915/16 was the reduction of gravity to curved space. The gravitational field, which embraced the inverse-square law of an attractive force between all bodies of matter, was

[23] At least, I have not been able to unearth any such document or documentation.

[24] For a further look at the context of these pictures, and how they have been manipulated and used various ways, see Topper and Vincent [199].

[25] Other interpretations propose that he was talking about cosmology; see Topper and Vincent [199]. I am grateful to Dwight Vincent for this interpretation of the equation.

[26] Due to the slip-shod nature of the squiggle, it is sometimes erroneously interpreted as the number 2.

expressed by the geometry of a curved space. $\mathbf{R}_{ik}=0$ expresses this geometrical interpretation of gravitational forces, with all the non-Euclidean geometry bound-up within the $\mathbf{R}_{ik}$ tensor. His quest was to extend this geometry to include non-gravitational fields, namely, electromagnetism, so that this unified theory too would be expressed in the form $\mathbf{R}_{ik}=0$.

A few years later he presented this problem as a metaphor (specifically an archi-tectural metaphor). To repeat: the task was to include the electromagnetic field, along with the gravitational field, into the equation. This may be symbolized this way: the energy of the electromagnetic field is abstractly written as **[e/m field energy]**, and this is added to the formulation $\mathbf{R}_{ik}=0$. This can be carried out by just putting the field on the right side of the equation, as follows: $\mathbf{R}_{ik}=$**[e/m field energy]**. This then would be a mathematical step toward encompassing the electromagnetic field within the gravitational field; in short, a step en route to unification. But this modified equa-tion (or the form of the equation) he felt was unsatisfactory. As Einstein wrote (and here, at last, is the metaphor): "But it [i.e., this modified equation] is similar to a building, one wing of which is made of fine marble (left part of the equation), but the other wing of which is built of low-grade wood (right side of equation)." This formu-lation was imperfect, for it did not fulfill his quest, as expressed a few sentences later: "The desire...[was] to include the gravitational field and the electromagnetic field in one unified formal picture."[27] In its present form the metaphorical equation was, abstractly put: **MARBLE = WOOD**. Einstein wanted to incorporate the wood <u>within</u> the marble. The result would be **MARBLE**$=\mathbf{R}_{ik}=0$. Can it be done? – that was the question in Einstein's lecture, succinctly expressed as $\mathbf{R}_{ik}=0$? If so, then there would be even more than ten equations embedded into $\mathbf{R}_{ik}=0$.

The goal to achieve this unity was the quest for rest of his life. In a commissioned article for <u>Scientific American</u> published in 1950 we still find the equation $\mathbf{R}_{ik}=0$ front and center. As my friend and colleague in Physics, Dwight Vincent, likes to say, $\mathbf{R}_{ik}=0$ was simply Einstein's favorite equation[28] – as he ever endeavored to turn the wood of the universe into pure marble, by reducing both the gravitational and electromagnetic fields to pure geometry.

By "pure geometry" I am referring to the mathematical unification of gravity and electromagnetism, not that Einstein believed in a mathematical world-view. He was at the start and remained to the end, a physicist, despite his changing opinions on the role of mathematics. As John Stachel has pointed out, Einstein no more thought his physics was being reduced to geometry, than Newton's inverse-square gravity law was reduced to a number for the parameter distance, or Maxwell's equations of electromagnetism to just vectors (notwithstanding Hertz's interpretation). There was always physics behind Einstein's equations, however esoteric the required mathematics was.[29]

---

[27] Einstein [47] [1936], p. 311. The essay is, "Physics and Reality," pp. 290–323, is reprinted from the *Journal of the Franklin Institute*, 221 (no. 3), March, 1936.

[28] At least in the last half of his scientific life, replacing, it seems, $\mathbf{E}=\mathbf{mc}^2$!

[29] Stachel [192], p. 244.

While on this specific point, it is worth recalling again the role of visualization or visual thinking in Einstein's pattern of thinking, which we delved into at the ends of Chaps. 12 and 15, above. The essence of the unification quest was a geometrical framework for electromagnetism and gravity, and this reveals the continuing role of visual imagery, since non-Euclidean geometry is the mathematics underlying physical fields. To summarize: we find a continuous thread from the "holy geometry book," to thinking in pictures in two thought experiments (riding a beam of light, and free falling in gravity), through the reality of the fields (electricity and magnetism) to the gravitational field as accounted for in the geometrical language of non-Euclidean geometry (visually as curved space), and finally toward a unification of electromagnetism and gravity under the structure of space-time. Such was the unflagging framework of Einstein's visual thinking over his life.

That the search for a unified field theory was indeed the topic of the Caltech lecture, and hence that this interpretation of the equation $R_{ik} = 0$ is correct, may be bolster by a report appearing in the journal Science the following October. The report pointed-out that "Professor Einstein's investigations of last winter" on the "Unitary Theory of Gravitation and Electricity" will be published shortly in Pasadena.[30] One may surmise that surely there was a strong incentive for Einstein to present those "investigations" in a public lecture during his California sojourn. In light of $R_{ik} = 0$? emblazoned on the blackboard in the photograph, I'm sure that he did.[31]

A final related question is this: How much of this world of tensor analysis was indeed understood by the experientialists present? Allan Sandage, who, as seen in the previous chapter inherited Hubble's job at Mt. Wilson, wrote this about Einstein's lecture in a personal communication: "I expect all this was over the astronomers' heads [since] they were all observers, not theoreticians."[32] Sandage probably was right.

* * *

In the same report on the projected publication of the unified field theory, a short outline of the theory was added, written by Einstein. After going over some details of the theory, he ended with the following curious comment: "The theory does not yet contain the conclusions of the quantum theory. It furnishes, however, clues to a natural development, from which we may anticipate further results in this direction."[33] To understand why this statement is curious we need to understand the deeper meaning of the quest for unity. For this we start with important methodological matters (the next chapter); and from there proceed to the quest, and the subsequent quarrel over the quanta, especially with Danish physicist Niels Bohr. The above curious statement will then commence the last Chapter in the book.

---

[30]Einstein [41]. We know that he also gave a lecture on the topic of a unified field theory to "thirty listeners" in a classroom in Pasadena on January 22, 1931, as reported in The New York Times, January 23, 1931, p. 17. This was a much larger audience than the small group in the library of the Observatory in the Pasadena offices.

[31]I have yet to find a date for this famous photo, but I surmise it should be near the January 22 of the previous footnote, since the topic is the same.

[32]Letter of May 1, 1998, also quoted in Topper and Vincent [199], p. 282.

[33]Einstein [41], p. 439.

Chapter 27
Exit, Mach; or, the Perils of Positivism

The ancient Greeks introduced an important philosophical or methodological concept into astronomy based on our limitations to direct access to the world of the heavens. It was part of what philosophers call epistemology, a word introduced in Chap. 17, and which we will use in the rest of this Chapter; the term refers to the process of acquiring knowledge of the world.[1] Obviously, we can touch and handle rocks; we can smell the roses; we can even, with the wave of a hand, feel seemingly invisible air. But we have only visual access to the Moon, Sun, stars and everything above. How therefore can we really know anything about them, beyond hypothesizing?[2] We can devise models and even test the models, but sometimes the same result comes from two different models. Without direct contact to the world above we can only deal with phenomena (what we see) not reality.[3] Thus arose the epistemological distinction between realism (our direct knowledge of the earthly world), and what we will call phenomenalism, for the appearances (phenomena) alone of the world. Although originally directed to our knowledge of the heavens, it easily was transferred to the larger epistemological question: How do we know anything?

This distinction (or dualism) between phenomenalism and realism has a long history from the late ancient world, through the Middle Ages, and into the Scientific Revolution.[4] By the seventeenth century, this dualism was centered primarily on whether the Earth really went around the Sun, or if the heliocentric model was just a simpler way of explaining the geocentric phenomena of the motions of the planets, Sun, and Moon. Those scientists who won out in the end (such as Copernicus,

---

[1]The word is derived from the Greek, *episteme*, meaning to know, from which we get empirical, experience, and importantly experiment.

[2]This again shows the significance of the invention of the spectroscope in the nineteenth century.

[3]The ancient Greeks often spoke of this as "saving the phenomena" or "saving the appearances."

[4]In the medieval world, this distinction played out in the Scholastic duality between what was called realism and nominalism, although their idea of realism was more compatible with what we would call idealism.

D.R. Topper, *How Einstein Created Relativity out of Physics and Astronomy*, Astrophysics and Space Science Library 394, DOI 10.1007/978-1-4614-4782-5_27,

Galileo, Kepler, and Newton) were all realists, whereas the phenomenalists at the time (often clergymen in the Catholic Church) were following a rear guard action. It would seem, therefore, that realism would triumph over phenomenalism as the next enlightened (eighteenth) century adopted Copernicus's model. In one sense it did, since the Sun-centered model was taught as being the real world. That world was held together by Newton's law of gravity, by no means a trivial achievement. Yet underpinning that law was the idea of action-at-a distance, a gravitational force acting and attracting all masses everywhere instantaneously. How could this be? How could forces travel across space? What was the real mechanism behind this phenomenon? Newton didn't say.[5] But, as quoted before (Chap. 4), near the very end of his <u>Principia</u>, he made the argument that it was "enough" that we know the law of gravity and how it accounts for the motions of the planets and the Earth's tides – without assigning a cause. Enough? Enough for whom?

This was blatant phenomenalism, and coming no less at the triumphal end of his otherwise realistic treatise. Blessed with Newton's authority, this phenomenalistic thread wove its way through the next two centuries, accompanying the realism of the Copernican model. A methodological approach to heat theory in the nineteenth century, for example, calculated heat flow without speculating on the physical nature of heat itself. Indeed, most of the science of thermodynamic, which was essentially born in that century, worked (and still works) within a phenomenalist framework, measuring pressure, volume, temperature, and such independently of the physical source of these variables.

As science progressed in the nineteenth century, parallel to the industrial revolution, the idea of science and engineering as engines of progress was amplified and science was further seen, even beyond what the eighteenth century Enlightenment envisioned, as a model for the intellectual, economic, and social worlds. The French philosopher Auguste Comte adopted the term positivism to designate his dream of a perfect society run by scientists and engineers, where theology and metaphysics would disappear and all knowledge would be positive by being grounded in empirical reality.

Positivism, in this sociological sense, and the accompanying appellation positivist entered the lexicon of scientific inquiry. In time, among some thinkers, it was synonymous with realism and the belief that science had hegemony over all human enterprises. The term positivism is still generally used this way today by journalists, popular science writers, and especially theologians (although the later usually do so pejoratively).

Paradoxically, in fact, from an epistemological viewpoint, the term instead became coupled with phenomenalism, because it was found that the goal of trying to ground science on empirical data alone, without preconceived ideas (or using the terminology of the time, being void of metaphysics), led to the realization that such

---

[5] In private, he agonized over it.

positive knowledge came at a price – the necessity to abandon the goal of a complete picture of the real world. This meaning of positivism, implying limitations to knowledge – the limitation of employing only empirical data without preconceived speculation – was, in fact, anathema to its popular (sociological) meaning. In short, by the late-nineteenth century, one meaning of the term positivism was part of the history of phenomenalism. Yet, concurrently, as it drifted into the last century, the Comptean (mainly social) version was synonymous with realism, although actually it was (and is) a variation of what more correctly is called scientism (the idea that science has supremacy over all forms of knowledge). In the rest of this book we will avoid this meaning, and use only the epistemological one.

This phenomenalist version – which is central to the topic here – was sometimes called critical positivism, and it produced several variations such as operationalism (from which we get the concept of an operational definition) and instrumentalism (such that knowledge is reduced to the reading of experimental instruments). The last version will be seen as particularly relevant Bohr's interpretation of quantum physics.

We need, however, one more background piece to this story, and this involves Ernst Mach, who once more enters the narrative, but from a slightly different point of view than before, and then he exits.

* * *

Mach was an important positivist who believed that all science is grounded on sensation, organized logically according to a principle of economy, from which concepts are formed by an inductive process, and further that legitimate science must eschew all else as mere metaphysical speculation. He deduced from this methodology that neither the aether nor atoms existed. He took this extreme skeptical view with him to the grave, still denying the realty of atoms when he died in 1916.

We have seen the influence of Mach on Einstein, with what Einstein called Mach's principle.[6] In addition, there was Mach's critique of Newton's absolute space and time as being sheer metaphysics.[7] The two were intertwined in Einstein's mind. But there was a wider context of Mach's impact on the early history of relativity, revealing further positivistic overtones. Mach's prerequisite to define clearly all variables in terms of their connection to the world of sensations is found in Einstein's 1905 relativity paper. Listen again to what I wrote in Part **II** about his definition of time.

> Einstein's discussion of time is presented in experiential language. Thus he writes: "If, for example, I say that 'the train arrives here at 7 o'clock,' that means, more or less, 'the pointing of the small hand of my watch to 7 and the arrival of the train are simultaneous events'." This sets the definition of local time as the time recorded by a local clock.

In addition to the role of what was called "railroad time" growing out of his work in the patent office, this emphasis on observers and events clearly betrayed the influence of one aspect of Mach's positivism on Einstein's early thought.

---

[6]Note, importantly, that Mach's anti-atomism had no influence on Einstein, since his non-relativity papers of 1905 were steeped in an atomistic worldview.

[7]Mach [134] [1912], Chap. 2, Sects. 6 and 7.

When Mach died, Einstein was asked to write an obituary.[8] He began by praising Mach as "a man who in our times had the greatest influence upon the <u>epistemological</u> [my emphasis] orientation of natural science, a man of rare independence of judgment." Einstein then made a plea for the general importance of epistemology to the progress of science and especially Mach's version in recent times. He emphasized how Mach's epistemological critique was central in questioning the "powerful authority" of "entrenched concepts" and, when necessary, dislodging them from science. The example he gave, not surprisingly, was Mach's critique of Newton, and Einstein copied long quotations from both Mach and Newton in the obituary. He even declared that Mach's work came close to articulating special relativity, and went so far as to say that Mach almost anticipated the equivalence principle.[9] The eulogy was published in 1917. The journal in which it was published received it on March 14, 1916. On March 20, 1916 Einstein's landmark summery paper on general relativity was received by the *Annalen*.[10] I mention the close proximity of these two events because they show that Einstein was writings both documents at the same time. This is interesting and ironic, in more than one way. The remainder of this Chapter will explain why. It begins with a brief overview of relativity from an epistemological viewpoint.

Despite Machian elements at the start of the theory of relativity, its history from 1905 through 1915/16 was also – in part, and from an epistemological viewpoint – a liberation or rejection of the strict Machian belief in the purging of speculation, guesswork, or hypotheses from proper science. A key step in this process, which (as seen) even Einstein initially balked at, was Minkowski's introduction of the concept of space-time into the mathematical formulation. But Einstein eventually came around to accepting that mathematical formalism, so much so that he acquiesced to using abstract tensor calculus to express the curvature of space. All this was certainly a flight from a physics based purely on sensations and experience. Therefore, at the same time that Einstein was holding vigorously to the specific concept of what he called Mach's principle, he was also bringing into relativity theory what Mach would pejoratively call metaphysics, and, as noted in a footnote above, using atomism in his other publications.

I am convinced that Einstein was consciously aware of this quasi-contradiction when he penned Mach's eulogy. For example, consider what he said in an address delivered only about two years later, in celebration of Planck's sixtieth birthday.

> The supreme task of the physicist is to arrive at those universal elementary laws from which the cosmos can be built up by pure deduction. There is no logical path to these laws; only intuition, resting on sympathetic understanding of experience, can reach them....Nobody who has really gone deeply into the matter will deny that in practice the world of phenomena uniquely determines the theoretical system, in spite of the fact that there is no logical bridge between phenomena and their theoretical principles....[11]

---

[8] *Einstein Papers*, Vol. 6, Doc. 29.
[9] Frankly, I find this a bit startling and erroneous.
[10] *Einstein Papers*, Vol. 6, Doc. 30.
[11] "Principles of Research," an address delivered before the Physical Society of Berlin, reprinted in Einstein [47] [1918], pp. 224–227; quotation on p. 226.

This statement expresses a radical epistemological departure from his early Machian positivism. There is no appeal to an inductive process from phenomena to scientific concepts, as dictated by Mach; instead, and contrarily, the physicist arrives at them intuitively – "there is no logical bridge between phenomena and their theoretical principles"[12] – and afterward they are tested by experiment. Clearly, Einstein's epistemological framework had dramatically changed from the quasi-positivism/ phenomenalism of 1905 into a variant of realism by around 1918, at the same time as he was praising Mach's purging of Newton's absolutes from physics. Surely the reason for the latter was because he still held to Mach's principle, which was based, in part, on rejecting Newton's absolute space.

What Einstein did not know while writing the obituary was that Mach not only was aware of this transformation of relativity, but – and in spite of his early support of the theory – he vigorously opposed what it had become by at least 1913, for in that year Mach wrote a strong rejection of relativity. This polemic, however, was not published until 1921, five years after his death.[13] In it Mach announced that he wanted "to cancel ... [his earlier] views of the relativity theory." He declared that he was seen "as the forerunner of relativity," but he now made it clear that he "assuredly disclaim[ed]" this role; in fact, he rejected the theory just as he rejected (and note the semantics) "the atomic doctrine." Mach continued with further religious allusions by lumping together relativity and atomism as part of "the present-day school, or church" which grow increasingly "more dogmatical."[14] He then promised to explain in a sequel why he held this new viewpoint, but such an explanation was never published.

Einstein became aware of Mach's rejection of relativity sometime in the 1920s, and it may have initiated his dismissal of Mach's principle that devolved over that decade, with (as seen in the previous Chapter) its final rejection in the early 1930s. A succinct summary of this new viewpoint appeared his Spencer lecture of 1933.[15] The topic was "the eternal antithesis between the two inseparable components of our knowledge, the empirical and the rational...." In Einstein's sweep through history, the Greeks bequeathed to western science the "admirable triumph of reasoning" based on Euclid's geometry. Then in the seventeenth century Kepler and Galileo provided the other component, empiricism. "Pure logical thinking cannot yield us any knowledge of the empirical world; all knowledge of reality starts from experience and ends in it," he wrote.[16] The second phrase of that sentence, it should

---

[12]Variations of this phrase he will repeat many times more in his writings over the rest of his life.
[13]Mach's rejection of relativity appeared in a preface to his book on optics, a posthumously published edition.
[14]Quoted in Holton [99], p. 248. This is a 1968 essay, "Mach, Einstein, and the Search for Reality," reprinted as Chap. 7, pp. 237–277, of Holton's book.
[15]The Herbert Spencer Lecture was delivered at Oxford on June 10, 1933: "On the Method of Theoretical Physics," reprinted in Einstein [47], pp. 270–276.
[16]Einstein [47] [1933], p. 271.

be noted, echoed the first sentence of Kant's Critique[17]– a book, recall, that Einstein was reading at Aarau while his fellow students were drinking beer. At this point in the argument, however, he departed from Mach. Unlike Mach, who saw concepts as fashioned by an inductive process from experience, Einstein said that concepts are "free inventions of the human intellect," not arrived at by a process of abstraction. He went on to pinpoint the source of this conviction; clearly, having followed Einstein's ideas about science this far, it comes as no revelation that he discovered the "erroneousness" of the abstraction notion when he worked on the general theory of relativity. As that theory came to fruition he realized that "the axiomatic basis of theoretical physics cannot be extracted from experience but must be freely invented," and further, he became "convinced that we can discover by means of purely mathematical constructions the concepts and the laws connecting them with each other, which furnish the key to the understanding of natural phenomena."[18] The years of struggling to find the correct tensor formation compatible with Newton's laws and more, no doubt triggered this transformation from his original quasi-Machian epistemological view. So radical was the change that he went on to write this astonishing declaration:

> Experience may suggest the appropriate mathematical concepts, but they most certainly cannot be deduced [really, induced[19]] from it. Experience remains, of course, the sole criterion of the physical utility of a mathematical construction. But the creative principle resides in mathematics. In a certain sense, therefore, I hold it true that pure thought can grasp reality, as the ancients dreamed.[20]

What a statement: "I hold it true that pure thought can grasp reality, as the ancients dreamed." Any positivist remnants still lingering in his mind were now flushed-out, as Einstein transformed himself into a realist.

As mentioned before, the aesthetic principle of covariance (which he had initially suspended) played a key role in the final formulation of general relativity and this had an impact on his thinking about scientific methodology. The Spencer lecture substantiates this. Furthermore, in the same year, he wrote this strong statement in a letter: "One should look for the mathematically most natural [aesthetic?] structures, without initially being bothered too much about the physical [empirical], as

---

[17]"That all our knowledge begins with experience there can be no doubt." This is the first sentence of Kant's, *Critique of Pure Reason*.

[18]Einstein [47] [1933], pp. 272–274.

[19]Einstein was actually speaking of induction here. In his writings he was often sloppy in his use of these logical terms, for he commonly used the terms induction and deduction interchangeably. I find this strange because he obviously knew the difference from his extensive reading in philosophy. Indeed, in an essay he was asked to write for the *London Times* in 1919 on "What is the Theory of Relativity?" he introduced a methodological distinction that he reiterated numerous times for the rest of his life. In it he distinguished between constructive theories that take a more inductive approach and are most common, and principle theories that start from first principles and are more deductive. The terminology (constructive- and principle-theories) was his own, and, as seen, the latter was the method of relativity. See *Einstein Papers*, Vol. 7, Doc. 25, which is reprinted from Einstein [47] [1919], pp. 227–232.

[20]Einstein [47] [1933], p. 274.

this [namely, the natural/aesthetic viewpoint] brought the desired result in gravitation theory."[21] Could the role of covariance be any clearer?

In the next decade in his autobiography, he summarized his flirting with Mach this way:

> It was Ernst Mach who, in his History of Mechanics, upset this dogmatic faith [in Newtonian mechanics as the foundation of physics]; this book exercised a profound influence upon me in this regard while I was a student. I see Mach's greatness in his incorruptible skepticism and independence; in my younger years, however, Mach's epistemological position also influenced me very greatly, a position that today appears to me to be essentially untenable. For he did not place in the correct light the essentially constructive and speculative nature of all thinking and more especially of scientific thinking....[22]

In short, during his epic journey from special to general relativity, Einstein's epistemological position shifted from phenomenalism to realism.[23] It was primarily within this latter frame of mind that he pursued the unification quest.

---

[21] Quoted in Van Dongen [205], p. 119. Letter to W. Mayer, February 23, 1933.

[22] Einstein [51] [1949], p. 21.

[23] I have used the term realist in a generic sense, which I have traced back to the ancient Greeks. For a closer reading of Einstein's realism as imbedded within the context of the wider philosophical debates in Einstein's time, see Howard [104].

Chapter 28
The Quest…and the Quarrel Over Quanta

In March of 1933, when Albert and Elsa left Caltech after their third sojourn abroad, they crossed the Atlantic and settled temporarily in Belgium, where he had a friendship with the King and Queen. The Einsteins lived in a small seaside town on the North Sea about seventy miles from Brussels, and were protected by security guards because the Nazis had put a price on his head. The Nazis emptied his bank account and ransacked his Berlin apartment several times, looting rugs, paintings, books,[1] and other sundry items. Fortunately a large collection of Einstein's scientific and personal papers were saved, taken to the French embassy, and smuggled out of the country by diplomatic pouch. How this happened, and which of Elsa's daughters was responsible for this act – Margot and her husband, Dimitri Marianoff, or Ilse and her husband, Rudolf Kayser[2] – is dependent on what source you read.[3] The Nazi's also raided Einstein's country cottage looking for weapons; they confiscated a breadknife.[4]

In Belgium, Einstein and Elsa were joined by his secretary, Helen Dukas, and his assistant, Walther Mayer, and eventually both daughters (Ilse and Margot) and their

---

[1] Marianoff [135] says they burned his books, p. 133.

[2] Kayser was the anonymous author of a biography of Einstein under the pseudonym, Anton Reiser (see Reiser [171]). Marianoff later also wrote a book on Einstein (Marianoff [135]).

[3] Marianoff [135] reports that on learning of the raid, he phoned Margot, directing her to take Einstein's papers to the French embassy (pp. 141–144). Isaacson [109], reports the story as true (p. 404), although he also says that Einstein "denounced" Marianoff's book (p. 559). Brian [15] reports that Dukas called Marianoff's book unreliable, although she conceded that many incidents were quite correct (p. 339). Hence, Brian repeats Marianoff's account of the Nazi raid in his book. Fölsing [65], however, credits Ilse and Kayser for saving the papers, library, and even shipping furniture to the United States (p. 666). Neffe [149], p. 287, echoes this, but does not cite a source. Pais [162] also credits Kayser for saving the papers (p. 528). Parker [164] says Ilse was in the apartment when the Nazis ransacked the place (Marianoff says Margot was there) and she was "scared out of her wits"; Parker thus credits Ilse and Kayser as saving the papers – but he provides on documentation (p. 234). Levenson [132] credits Margot for sending Einstein's important documents to the French embassy, but he too cites no source (p. 419). It appears that many secondary sources are merely copying each other.

[4] The breadknife story is reported by Marianoff (p. 144).

D.R. Topper, *How Einstein Created Relativity out of Physics and Astronomy*, Astrophysics and Space Science Library 394, DOI 10.1007/978-1-4614-4782-5_28,
© Springer Science+Business Media New York 2013

husbands. He also met Lemaître again, and together they organized some seminars in Brussels.[5] During this Belgium stopover, Einstein also made a side trip to Britain to deliver several lectures,[6] and a bittersweet trip to Switzerland to visit with his younger son, Eduard – it would be the last time they met.[7]

In October, the Einstein's, Dukas, and Mayer arrived in the United States and proceeded to the university town of Princeton, New Jersey, where Einstein took up his new post at the Institute for Advanced Study. Ilse and Margot remained in Europe. The following year Ilse died in Paris, and Margot (along with her husband) moved to Princeton. Einstein purchased a modest house at 112 Mercer Street – modest, yet large enough to accommodate he and Elsa, and Helen Dukas. Sadly, a year later Elsa died, and Dukas assumed the homemaker duties in addition to her secretarial tasks. After Margot and Marianoff separated (1934), she moved-in; in 1939 Einstein's sister, Maja, left her husband behind in Europe and she moved-in, too. Later, when his sister lay sick and dying,[8] Einstein read to her works of literature and history of science.[9]

* * *

One theme of this is book is the influence of other contemporary scientists on Einstein, despite his intransigence and originality. In his personal life he could be extraordinarily friendly with a joyous and affecting laugh, and therefore was often well-liked by those he met, forming deep friendships with key physicists of his time with whom he kept-up extensive correspondence.[10] But his relationships with members of his immediate family were usually (and unusually) strained. Crassly put, Einstein was often a lousy husband, and not much better as a father.[11] True, his efforts for many social causes revealed a real love of humanity, but ultimately he preferred to be alone and therefore he could be difficult to live with in close quarters.

---

[5]Farrell [61], pp. 118–119. Their last meeting took place in Princeton in 1935, where Lemaître was lecturing for a semester at the Institute, but unfortunately there seems to be no documentation of their interaction. Farrell, p. 119.

[6]One was the previously mentioned Herbert Spencer Lecture delivered at Oxford on June 10, 1933, reprinted in Einstein [47], pp. 270–276. Also, the lecture, "Notes on the Origin of the General Theory of Relativity," was delivered at the University of Glasgow, June 20, 1933, reprinted on pp. 285–290.

[7]Pais [162], pp. 450–451. According to Neffe [149], p. 199, there is no documentation of what transpired during their meeting.

[8]Maja died there in 1951. Helen Dukas outlived Einstein (d. 1955), remaining in the house until her death in 1982. Margot, who inherited the house, died in 1986. She also received $20,000 from Einstein's will, as did Dukas, which was more than his sons' inheritance (Hans received $10,000 and Eduard $15,000). Neffe [149], p. 191; Michelmore [141], p. 258.

[9]Einstein [55] [1948], pp. 105–107 (letter to Solovine, November 25, 1948); Isaacson, p. 518.

[10]This is relentlessly witnessed by historians as the volumes of the *Einstein Papers* are sporadically published. I, at least, find the correspondence overwhelming.

[11]Neffe [149], Chaps. 6 and 10. Neffe, in particular, seems to focus heavily on Einstein's personal flaws. The most castigating viewpoint is Highfield and Carter [94], *passim*. For a more positive reading of Einstein's character, see Frank [67], who knew him personally.

In his scientific life, in some ways his "closest" scientific colleagues were dead: Galileo and Newton, or Faraday and Maxwell – these being perhaps more important than his living colleagues. Indeed, a key theme of this book is Einstein's penchant for delving into unresolved problems of past scientists. In so doing he probed deeper than others into the core of ideas, assumptions, and taken-for-granted postulates of physics. It is not surprising, therefore, that on the walls of his study in his apartment in Berlin and later in his home in Princeton were pictures of three scientists: Newton, Faraday, and Maxwell, all of whom, along with Galileo, formed a quartet of deceased men whose ideas he both used and rejected.[12]

In addition to being accompanied by pictures of past scientists on the wall, Einstein, in his domestic life during much of the Princeton years, was surrounded by women. In his later years he mellowed in his personal relationships. Indeed, he always enjoyed the company of woman outside his immediate family,[13] although his attitude toward them was, let us say, more Victorian than progressive – unlike his physics.

As perceived by his scientific colleagues during the Princeton years, Einstein was also seen as less than progressive: he passionately, stubbornly, and almost solely[14] pursued the unification quest; and he did not give ground in his intellectual quarrel with other physicists over what he saw as the incompleteness of quantum physics. These dual efforts, which were seen as rearguard obstinacy by most scientists, made

---

[12]At least, there is a general consensus it was these three. Hoffmann [97], pp. 46–47, names these three in Berlin (reproducing the images of Faraday and Maxwell on p. 46) but says the image of Newton was lost in the Nazi raid of 1933. In Sugimoto [194], p. 102, there is a photo of Einstein sitting in his Berlin study with Newton's picture clearly on the wall, and which is reproduced larger on the same page. The same photo is in Renn (ed.), [172], Volume 1, p. 421. Marianoff [135], p. 1, first mentions "a large framed picture of Michael Faraday" in the Berlin apartment Library; then, on page 205, he says that in Einstein's study in Princeton there were three pictures that came from Germany with the furniture: Newton, Maxwell, and Faraday, thus contradicting Hoffman. He also said there were no other pictures in the room. Reiser [171], pp. 193–194, says the Berlin study had pictures of Faraday, Maxwell, and Schopenhauer, the latter being one of Einstein's favorite philosophers, along with Kant, Hume, and Spinoza. Bucky [22], pp. 51–52, also names these three in Berlin. Isaacson [109], p. 438 (but with no citation), says that at Princeton there were pictures of Newton, Faraday, Maxwell, and later Gandhi; whereas Bucky [22], p. 42, says there were only pictures of Faraday, Maxwell, and Gandhi in Princeton, which seems to support Hoffmann on the missing Newton picture. Clark [26], p. 643, names the same three, without a citation. Further evidence is supplied by the physicist, R. S. Shankland [182], pp. 54 and 57, who visited Einstein in his study in 1952 and 1954, and mentions only pictures of Maxwell and Faraday, and science historian I. B. Cohen [27], p. 69, who visited Einstein two week before he died, and mentions the same two pictures. These last reports may not necessarily eliminate the picture of Gandhi, since as science historians they may only have been interested in scientists on the wall.

[13]For a brief overview see Highfield [93].

[14]Initially there were several unification attempts by Arthur Eddington, Theodor Kaluza (see below), and others, but eventually Einstein (with some collaborators, also below) was alone in the quest.

him increasingly isolated from the mainstream of science; the most egregious derision was his being seen as a stubborn old fool. A young physicist visiting the Institute
once said he was "completely cuckoo."[15]

There were actually a number of reasons for this isolation. For one, the unification
quest did not, in time, seem to bear fruit. When he began the quest around 1920
there were only three fundamental particles in nature: electrons, protons, and photons. Also around the same time the German mathematical-physicist, Theodor
Kaluza, put forward the idea of unifying electricity and gravity by extending relativity to a fifth-dimension. Why not: if gravity is the fourth-dimension, why cannot
electromagnetism be the next one? Einstein liked this idea.[16] It was potentially a
way of incorporating the field model of electromagnetism into the field of gravity,
within the framework of non-Euclidean geometry. Later Kaluza's worked was
extended by the Swedish physicist, Oscar Klein, and it became known as the Kaluza-
Klein theory.[17] Einstein picked up this idea and tried to make it work. The calculations were formidable and he was aided in this by his assistant Walther Mayer, who,
as mentioned before, was nicknamed "Einstein's calculator."

Mayer, who came with Einstein to Princeton,[18] was the first in a line of many
assistants working with him at the Institute on the celebrated quest.[19] From the
mid-1920s, when Einstein was in his mid-40s, and over the next twenty years, he
tried various mathematical models to achieve unification, coming back about every
five years to the some variation of the Kaluza-Klein model, until he eventually
gave up on it entirely.[20] The five-dimensional model had seemingly the same conceptual attraction for him that Mach's principle had. When Einstein liked an idea,
he was reluctant to give it up. It is interesting that in 1901, in a letter to Grossmann,
he wrote, "It is a glorious feeling to perceive the unity of a complex of phenomena

---

[15]This physicist was none other than a young J. Robert Oppenheimer, who later directed the
scientific part of the Manhattan Project for building the bomb. The quotation is from a letter to his
brother, Frank, in 1935, in Oppenheimer [160], pp.189–191. After the war Oppenheimer became
the Director of the Institute (1947–1966) and developed an affable relationship with Einstein,
declaring in a memoir: "Just being with him was wonderful." Oppenheimer [159], p. 47.

[16]Pais [162], p. 330.

[17]The complex details of this story, and Klein's interactions with Einstein, are found in Halpern
[86].

[18]Part of the negotiations for the position at the Institute included Einstein's stubborn insistence
upon a salaried position for Mayer. See Pais [162], Chap. 29.

[19]Others included Bannesh Hoffmann (1906–1986), Leopold Infeld (1898–1968), Peter Bergmann
(1915–2002), Valentin (Valya) Bargmann (1908–1989), Ernst G. Straus (1922–1983), and others.
His last was a woman, Bruria Kaufman (1918–2010), an Israeli physicist. Many were part of the
brain-drain from Germany. Interestingly, Bergmann wrote the first English textbook on general
relativity. See especially the discussions among Hoffmann, Bergmann, Bargmann, and Strauss in
Woolf [215], Section XI. Also see Renn (ed.), [172], Volume 2, p. 143, and Van Dongen [205], pp.
140–142 and 146–148.

[20]Pais [162], p. 342. Halpern [86], p. 401, sets the date for the final rejection as 1942. Van Dongen
[205], Chap. 6, sets the final date as 1943, p. 153.

which appear as completely separate entities to direct sensory observation." Little did he know what sort of unity he would pursue later in his life. Also worthy of note in the same letter is a reference to his own stubbornness: "God created the donkey and gave him a thick hide." So, maybe Einstein was less a cuckoo and more a donkey – at least it is his own nickname.[21]

By the mid-1940s Einstein may have given-up on the Kaluza-Klein theory but not the quest, which he continued to the end. As seen above, as late as 1950, in the article for <u>Scientific American</u>, he was still plugging away at his favorite unification equation, $R_{ik}=0$. Also, late in 1954 he completed what would be his last scientific paper: he said it was "the logically simplest relativistic field theory which is at all possible."[22] Several months later, on the nightstand next to his hospital bed on the morning of April 18, 1955, pages of loose papers were gathered-up by the night nurse who was putting together his belonging, since he died of a ruptured aortic aneurysm at 1:15 a.m. The pages contained yet another, further, quest to unify the forces of nature. The donkey in him persevered to the very end.[23]

This leads us to another reason for his scientific isolation later in life.

By the 1930s there were many more fundamental particles found in nature than the previous three: first the neutron, then the neutrino,[24] next the muon (of which it was soon found there were two, with positive and negative charges), followed by the pion (again two, with positive and negative charges), and then a barrage of particles, the kaon, lambda, sigma, and more, produced by high energy accelerators

---

[21] *Einstein Papers*, Vol. 1, Doc. 100. Letter to Grossmann, 14 April, 1901. It should be pointed-out that the context of his stubbornness was not a scientific topic; rather, his lack of acquiring a job and his not giving up in trying to procure one.

[22] Einstein, "Relativistic Theory of the Non-Symmetric Field," published as Appendix II, in the fifth edition (1956) of the Princeton lectures, *The Meaning of Relativity*, quotation on pp. 163–164. His preface to this edition is dated December 1954, and in it he notes that the paper was written "in collaboration with" his assistant, Bruria Kaufman, his last collaborator. For a recent and more detailed technical study of Einstein's quest, see Van Dongen [205].

[23] Einstein requested there be no physical memorial to him; he wanted no grave, and that his ashes be scattered in an unknown place. His ashes were scattered by Otto Nathan, who was a close friend of Einstein and the sole executor of his will. They met when Nathan taught economics at Princeton University (1933–1935), and they remained life-long friends. Their friendship was based, in part, on their mutual leftist political views; in fact, Nathan was a harassed during the McCarthy witch hunt for such views. Most sources state that Nathan was responsible for scattering Einstein's ashes in an unknown place. But Michelmore, 1962, said it was a nearby river (p. 262). As noted in my annotation to Michelmore's book in my bibliography, the book was based in part on interviews with Einstein's son, Hans, as well as Helen Dukas and Nathan, the two being the co-trustees of Einstein's estate. The latter fact brings to mind another point that should be briefly made on their roles in blocking the publication of documents that revealed the darker side of Einstein. Using litigation, Dukas and Nathan repeatedly delayed publication of the *Einstein Papers* and other works revealing the less-heroic side of Einstein. Dukas died in 1982, and it is no accident that when Nathan died in 1987, later in the year the first volume of the *Papers* finally was published. For more on this see Stachel [192], pp. 95–103, and Highfield and Carter [94], pp. 243–285.

[24] At least the postulation of the neutrino took place. The measurement of its actual existence was not made until the mid-1950s. A short version of this story is in Topper [198], pp. 85–87.

(commonly known as atom smashers). Along with the proliferation of this "zoo" of particles came two more forces in nature within the nucleus of the atom: the strong nuclear force, and a weaker radioactive force (appropriately, although not very creatively, called the "weak force"). Now there were four forces in nature. Does this not mean that for a complete unified theory, all four forces should be unified? Apparently the answer among the majority of physicists was "yes," for by focusing only on the unification of gravity and electricity, Einstein's quest was seen as increasingly extraneous and irrelevant to the mainstream.

It is also significant that Einstein ignored the development of nuclear physics,[25] a major new field of physics of the twentieth century, which surely was another factor for his isolation. For some readers it comes as a shock to learn that Einstein never wrote a paper on nuclear physics, especially since the common view of the equation $E = mc^2$ is associated with the bomb and nuclear energy. Indeed, the Time magazine cover from July 1, 1946, depicted Einstein in the foreground, with a nuclear bomb-blast behind him, and $E = mc^2$ inscribed within the mushroom cloud. Yet anyone who has studied the development of the building of the bomb at Los Alamos, New Mexico from 1942 to the first test detonation in 1945, realizes that it was mainly a technological accomplishment, based on nuclear science (such as the discovery of the possibility of a nuclear chain reaction) – and, importantly, except for that little equation, Einstein contributed next to nothing to the endeavor.[26] From a socio-political perspective he was isolated from the project. It is true that in 1939 he was involved in writing a letter to President Roosevelt warning him of the possibility of the German's developing such a weapon, but otherwise he was cut-off from the project by the US military and the FBI.[27] Because of his previous associations in Europe with various left-wing causes, coupled with American paranoia over Communism, Einstein was deemed a security risk. Indeed the FBI amassed a large file on him during his years in the United States, not only for his previous "subversive" activities in Europe, but also his outspokenness on social and political matters in his adopted country, such as human rights (he was particularly vocal on the "Negro question," namely the treatment of African-Americans), and later was vocal in his objection to the anti-Communist hearings during the McCarthy era.[28] In the end, I suspect, it did not matter that he was not involved with the doings at Los Alamos, for he probably knew little of nuclear physics anyway.

As noted before, by the mid-1920s Einstein ceased making contributions to quantum physics, and this brings us to the last reason for his isolation – namely, his

---

[25] Holton [100], p. 166.

[26] The caption to the Time picture is bizarre. It reads: "Cosmoclast Einstein," with the subtitle, "All matter is speed and flame." The neologism "cosmoclast" is apparently a combination of cosmologist and iconoclast. The subtitle, presumably, is a cryptic reference to $E = mc^2$ within the context of the bomb.

[27] For a recent account pointing to some involvement by Einstein, see Schweber [180], pp. 42–62, especially p. 51.

[28] Green [81], Introduction.

disengagement from the quantum theory, and, most importantly, his strong objections to its methodological interpretation. Quantum physics was center stage and making what was seen as fundamental progress in the 1920s. Einstein himself, as mentioned, was still part of this development,[29] making his last contribution in mid-decade.[30] By his mid-forties, however, he exclusively devoted his scientific efforts to the unification quest, just about the time that the theoretical framework of quantum physics shifted toward an interpretation of the theory that he was convinced was, in his word, "incomplete."

A constant that runs throughout this account of his work is a contrarian or non-conformist attitude: from the origin of special relativity in his youth, through a resistance to where quantum physics was going in his middle age, to an obstinate pursuit of the unified theory for the remainder of his life.

* * *

To understand Einstein's quarrel with the quanta it helps to find its roots. This takes us back at least to 1905 and his first paper where he put forth his hypothesis of applying the idea of a quantum of energy as a fundamental entity in physics. This work was briefly mentioned in Part **II** in the context of that miracle year. But, the history of quantum physics and even Einstein's contribution to it is not the topic of this book on relativity. The history of quantum physics is far beyond the scope set out here: the story involves an extraordinarily complex historical narrative with numerous players. Hence, I merely present what is deemed necessary for comprehending the essence of his critique of the "incompleteness" of quantum physics.

One physicist at the center of this subject was Einstein's personal friend but intellectual antagonist, Niels Bohr[31] famous for making the first workable model of the atom in 1913, for which he received a Nobel Prize. Their disagreement from the late-1920s into the early-1930s has been called by historians of physics the Bohr-Einstein debate, although it was more a dispute or occasional bantering than a formal debate (see Photo 28.1).[32] At the center was a difference in interpretation of the statistical nature of the world opened-up with the discovery of the atomic realm, and the interpretation of the experimental results regarding the wave-like and particle-like nature of, first, light and then, later, matter itself.

In Einstein's contribution to the quantum theory from the 1900s right into the mid-1920s (at first alone, and then in collaboration with others), he drew, in part, on and continued the history of the use of statistical methods in many areas of physics and other sciences, a process that went back into the seventeenth century but which made

---

[29] Stachel [192], p. 385.

[30] It was not a minor contribution, either. In 1924–1925, in collaboration with the Indian physicist, Satyendra Nath Bose, they predicted something called the Bose-Einstein condensate (later experimentally found), employing, and accordingly introducing, what is called Bose-Einstein statistics.

[31] (1885–1962).

[32] Klein [117]. Klein, however, refers to it as a "dialogue." See also, Brush [17], pp. 414–419. The literature on this is extensive.

Photo 28.1   Niels Bohr and Einstein probably at Paul Ehrenfest's home in Leiden, circa 1925–1930. Restoration of original negative and print by William R. Whipple. Photograph by Paul Ehrenfest, courtesy of Emilio Segre Visual Archives

major progress in the nineteenth century.[33] Einstein certainly had no qualms employing statistics to account for phenomena in nature, such as the law of gases.[34] A short but important paper that provides crucial insights on this issue is Martin Klein's study of Einstein's pre-1905 papers using statistics (1902–1904).[35] He reveals that in Einstein's application of probability theory to mechanical problems "the mechanics gradually become less prominent while the purely statistical features of the theory take on more importance."[36] Klein contends that Einstein was a pioneer in the application of statistics to particular problems in physics, which casts further light on his central role in the development of the quantum theory, from the beginning to at least the mid-1920s.[37] This then brings us to the crucial question around which his debate with Bohr circled: "What does the use of statistics in physics tell us about the physical world?"

Fortunately, there is a conceptual way of getting to the heart of the matter without plowing through the mathematical and physical details of quantum physics; namely, by borrowing two concepts (one of which was already introduced) from philosophy – these being, epistemology and ontology. Epistemology, as seen, is the study of how we know what we know; it looks at the process of deriving knowledge about the world or reality. This was crucial to the history of what we called

---

[33] In addition to its application in physics, statistics was used in the social sciences, and in general to experimental error; think of the Gaussian distribution, or the so-called bell curve. Brush [17], pp. 399–403, emphasizes the rise of statistics in nineteenth physics, such as around the problem of irreversibility in thermodynamics.

[34] Brush [16], p. 92.

[35] Klein [118].

[36] Klein [118], p. 115.

[37] This nicely dovetails with Kuhn's thesis, 1987.

phenomenalism and its later variant dubbed positivism. Such a study led to the possibility of limitations to our knowledge, therefore setting bounds to our understanding of the external world. This we have seen. The new subject is ontology,[38] which is the study of what prosaically is call reality, or what the world is – that is, the being of the world. In short, the two terms engage with <u>how</u> we know (epistemology) and <u>what</u> we know (ontology) about the external world.[39]

At this juncture it is worthwhile to keep in mind the sentence from Einstein's autobiography that was quoted several times before, when he recalled his pre-teenage scientific revelation shortly after his rejection of a religious mind-set of the world: "Beyond the self there was this huge world, which exists independently of us human beings and which stands before us like a great, eternal riddle, at least partially accessible to our inspection and thinking." Two truths were revealed: the existence of an external world, and our (partial) access to that reality. The limitation (or possible limitation) implied in the last phrase, was for Einstein (to use our philosophical terminology) an epistemological limit not an ontological one. The world's reality was not affected by, say, statistical factors that come into play when we perform experiments. Deriving phenomenological parameters such as the pressure or temperature of a gas using statistical methods, had no bearing on the real makeup of the gas, which was composed of real molecules bouncing around like billiard balls. I think this is what he meant, around 1949, in the following unpublished statement. "In truth, I never believed that the foundation[40] of physics could consist of laws of a statistical nature."[41] Statistics was part of the epistemology of scientific knowledge, it had no bearing on the ontology of the world.

In addition to the question of the statistical nature of quantum physics, there arose the other fundamental problem to which Einstein contributed from the start. His introduction of the quantum of energy for light revived Newton's idea of light as having a particulate nature, as opposed to the wave model that spread through the nineteenth century and which we know was linked to the development of field theory that Einstein drew on for his relativity theory. Resistance to the concept of a light particle was based mainly on its contradiction with the phenomenon of interference, something that was interpreted as explainable only on a wave model. As see, even Millikan, who performed an experiment that was crucial for confirming Einstein's quantum equation, initially set-up the experiment to disprove the particle model, and that the term photon for the particle of light was not coined until the mid-1920s shows the resistance to the concept by most physicists.[42] This included Bohr, who in the early-1920s held to the position that the quantum light model came

---

[38] Ontology is derived from the Greek word *on*, meaning to be or to exist.

[39] The reader may wish to compare my approach to these methodological matters with that of Van Dongen [205], Chap. 2.

[40] *Grundlage*, in German.

[41] From an unpublished reply to Max Born's essay in Schilpp (ed.) [12]; quoted in, Stachel [192], p. 390. See Born [12].

[42] Brush [21], p. 223 note 52 and p. 227.

into play only during the interaction of light with matter, not light alone in a vacuum, for which Maxwell's field (that is, wave) equations applied.[43]

But with the eventual recognition of the photon as an independent reality within the world of fundamental particles in the late-1920s,[44] the reality emerged that light exhibited a dual nature. Einstein, in fact, was instrumental in articulating this wave-particle duality in physics as far back as 1909,[45] since he was almost alone in initially accepting the reality of the quantum of light. In his autobiography he recalled that as far back as his student years he was aware of "the disturbing dualism"[46] between the particle masses in Newton's theory and the waves of Maxwell's field, a dualism he hoped to be able to eliminate.[47]

As the photon's reality was gradually acknowledged, a French physics student, Louis De Broglie,[48] put forth an even more radical idea in 1923 when he postulated that matter too should be dualistic, in symmetry with light. Essentially an aesthetic argument – accompanied, appropriately, with equations – this avant-garde notion was initially hypothesized in a draft of his Ph.D. thesis. One of his examiners was Langevin, who sent a copy to Einstein, for his appraisal. Einstein, in reply, called it "very interesting" and went on to declare that it was "a first feeble ray of light on this worst of our physics enigmas." I do not interpret his use of the word feeble pejoratively, but rather that Einstein felt De Broglie was making a beginning, a significant beginning, however small – since no one else was casting much light on the duality enigma. Einstein himself, it turned-out, was pursuing a similar line of thought in his work on quantum gas theory. De Broglie's thesis was accepted, published in 1924, and so physicists were faced with the concept of matter, too, having a wave nature. But was there really something like matter-waves?

As seen, the particle nature of matter was recognized in 1897 with J. J. Thomson's discovery of the electron. Interestingly, and even ironically, De Broglie's idea was tested by G. P. Thomson,[49] J. J. Thomson's son. Sending a beam of electrons through a thin metal foil in a vacuum, he produced interference patterns. Yes, electrons exhibited wave-like behavior. G. P.'s experiment revealing the existence of matter-waves was performed in 1927, thirty years after his father's confirmation of the particle nature of matter.[50]

All of nature, matter and light, seemed to be dualistic, presenting both wave-like and particle-like behavior. Physicists were now confronted (or perhaps better said,

---

[43] Klein [117], p. 13.

[44] This transpired mainly through a series of further experiments that confirmed the photon. The details, however, are beyond the scope of this book.

[45] Klein [117], pp. 4–6; Stachel [192], p. 379.

[46] *der störente Dualismus*, in German.

[47] Einstein [51] [1949], p. 35.

[48] (1892–1987).

[49] George Paget Thomson (1892–1975).

[50] Pais [162], pp. 436–437. De Broglie received a Noble Prize in 1929 and G. P. Thomson in 1937. There were other experiments confirming the wave nature of matter.

confounded) with deeper questions. How can these opposites be reconciled? What does this dualism, along with the need for a statistical understanding of phenomena, tell us about the world? Einstein is reported to have said in a lecture in 1927: "What nature demands from us is not a quantum [particle] theory or a wave theory; rather, nature demands from us a synthesis of these two views which thus far has exceeded the mental powers of physicists."[51] This limitation he saw as epistemological; for him, this dualism was about our limited knowledge of reality, as was the role of statistics. Nature itself was neither particle-like nor wave-like. There was a deeper reality presently beyond our comprehension. A real or complete theory would produce "a synthesis of the two views," Einstein believed.

* * *

Bohr and Einstein first met in Berlin in 1920 when Bohr presented a guest lecture. We know something about their meeting from a follow-up letter from Einstein, where he spoke of his "joy" at the "mere presence" of Bohr. In his reply, Bohr called the meeting "one of the greatest experiences ever."[52] Much later, in his essay in the Schilpp collection, Bohr recalled that they discussed the topic of statistics in quantum physics, and the ensuing problem of causality, for if nature is fundamentally (read: ontologically) statistical, then strict cause and effect determinism is not viable. This "formed the theme of our conversation" he wrote, and noted that Einstein was already reluctant to abandon "continuity and causality."[53] Despite their obvious mutual admiration at that initial meeting, Einstein's realism seemed fixed from the start, and Bohr's challenge was to pry away at this rigidity.

A major attempt by Bohr to interpret the wave-particle dilemma was presented in a lecture in 1927, just as the duality of matter and light was being confirmed. To physicists gathered at a conference in Como, Italy, he put forth what he called the complementarity principle. This principle he conceived as a way around the duality problem. The difference between waves and particles did not necessarily contradict each other if one kept in mind the role of experiments. Each experiment brought-out different aspects of the subatomic world. Here was how he later expressed this idea: "...evidence obtained under different experimental conditions cannot be comprehended within a single picture, but must be regarded as complementary [Bohr's emphasis] in the sense that only the totality of the phenomena exhausts the possible information about the objects."[54] The role of empirical induction in this process betrayed the phenomenalism or positivism toward which Bohr was groping.

Einstein was not present at the Como conference, but they did meet the next month at a conference in Brussels. Bohr's explanation and justification of his complementarity principle did not budge Einstein from his realist position. As well,

---

[51] Quoted in Pais [162], p. 443.
[52] Quoted in Pais [163], pp. 227–228.
[53] Bohr, in Schilpp (ed.) [10], pp. 205–206.
[54] Bohr, in Schilpp (ed.) [10], p. 210.

Bohr quotes Einstein as having invoked the Deity in defending his determinism by saying that the world would be statistical only "if the dear Lord plays at dice."[55] Much is often made of Einstein's appeal to God on this statistical problem. According to Arnold Sommerfeld, Einstein increasingly made reference to God in the 1920s, especially when a new idea "appeared to him [as] arbitrary or forced." Sommerfeld then quotes Einstein as saying, "God doesn't do anything like that."[56] It may be worthwhile to recall the Prague years, and Einstein's possible re-interest in Spinoza, for Sommerfeld also quotes from a New York Times article of 1929, where Einstein was asked by a Rabbi if he believed in God, and Einstein is quoted as saying: "I believe in Spinoza's God, who reveals himself in the lawful harmony of all that exists, but not in a God who concerns himself with the fate and the doings of mankind."[57] Whether or not this corresponds to Spinoza's actual theological position I leave to scholars of Spinoza, but it does tell us two things about Einstein: he did not believe in a traditional transcendent God, and he was not a pantheist; the latter follows from Einstein's idea of God revealing himself, an act that implies a non-identification of God and Nature – something anathema in a pantheistic worldview. Of course, I also have a hunch that some of Einstein's expressions around the Deity were sometimes made in jest.[58]

Leaving theology briefly and returning to quantum physics, consider this sentence from a late letter where Einstein made a noteworthy argument about quantum physics. "The sore point[59] lies less in the renunciation of causality than in the renunciation of the representation of a reality thought of as independent of observation."[60] The issue of a world independent of human perception was fundamental to his belief and instinct – "Beyond the self there was this huge world, which exists independently of us human beings..." – and the loss of this separation was more difficult to fathom than allowing statistics to play a role (even perhaps ontologically?) in physics. Returning to theology, now coupled with the quantum problem, Einstein once wrote to Max Born (not Bohr!): "You believe in the God who plays dice, and I in complete law and order in a world

---

[55]Bohr, in Schilpp (ed.) [10], p. 218: "...ob der liebe Gott würfelt." About the same time, in a letter to Max Born, December 4, 1926, he wrote a now-famous statement: "Quantum mechanics is certainly imposing. But an inner voice tells me that it is not yet the real thing. The theory says a lot, but does not really bring us any closer to the secrets of the 'old one.' I, at any rate, am convinced that *He is* not playing at dice." In Einstein [56] [1926], p. 88, emphasis his.

[56]Sommerfeld, in Schilpp (ed.) [189], p. 103.

[57]I am using the quotation from Isaacson [109], pp. 388–389, who is quoting from the New York Times, April 25, 1929; cited in Isaacson, p. 617, note 9.

[58]It is of some relevance that Einstein's son-in-law, Rudolf Kayser, who wrote a biography of Einstein (a.k.a. Reiser [171]), also wrote a book on Spinoza (see Kayser [114]), for which Einstein wrote the Introduction. In it Einstein discussed only the psycho-social world of Spinoza and compared it to the present post-Second World War situation, making no mention of philosophical or theological matters. I'm not sure what to make of this.

[59]*Der wunde Punkt*, in German.

[60]Letter to Georg Jaffe, January 19, 1954, quoted in Stachel [192], p. 390.

which <u>objectively exists</u> [my emphasis], and which I, in a wildly speculative way, am trying to capture."[61] The independent world precluded any statistical ontology.

Einstein grew more adamant in his realism as Bohr drifted further and further into phenomenalism. In a 1936 essay "Physics and Reality," Einstein reiterated what he saw as the incompleteness of quantum physics. He began by acknowledging the experimental confirmation of the theory. "Probably never before has a theory been evolved which has given a key to the interpretation and calculation of such a hetero-geneous group of phenomena of experience as has quantum theory." But he added an important caveat:

> In spite of this, however, I believe that the theory is apt to beguile us into error in our search for a uniform basis for physics, because, in my <u>belief</u> [my emphasis], it is an <u>incomplete</u> [Einstein's emphasis] representation of real things…. The incompleteness of the represen-tation leads necessarily to the statistical nature (incompleteness) of the laws.[62]

That this really was Einstein's <u>belief</u> is made clear a bit later:

> To believe this [that is, quantum physics] is logically possible without contradiction; but, it is so very contrary to my scientific <u>instinct</u> [my emphasis] that I cannot forego the search for a more <u>complete</u> [Einstein's emphasis] conception.[63]

Bohr was neither enamored nor swayed by Einstein's "belief" and "instinct," and he responded this way: "[I]n quantum mechanics,[64] we are not dealing with an arbi-trary renunciation of a more detailed analysis of atomic phenomena, but with a recognition that such an analysis is <u>in principle</u> [Bohr's emphasis] excluded."[65] This is a famous and often quoted statement.[66] Let me therefore repeat what he said, but by interjecting my terminology: "[I]n quantum mechanics, we are not dealing with an arbitrary renunciation of a more detailed analysis of atomic phenomena [that is, the incompleteness is not due to an epistemological limit because of, say, the need for statistics in a complex system], but with a recognition that such an analysis is <u>in principle</u> excluded [that is, the statistical and dualistic description is an ontological fact of nature; there is no further or deeper reality to be found]." Bohr was edging ever closer to becoming a true phenomenalist/positivist.

There is a striking story told by Bohr's assistant that took place in 1939, when Bohr was at the Institute in Princeton for a few months. The assistant, who accom-

---

[61] Letter of 1944, in Einstein [56] [1944], p.146.

[62] Reprinted in, Einstein [47] [1936], pp. 290–323, quotation from pp. 315–316.

[63] Einstein [47] [1936], p. 318.

[64] In the late 1920s the terms quantum physics and quantum theory were often replaced by quantum mechanics, which usually referred to the transformation of the theory by Erwin Schrödinger, Werner Heisenberg, and others – needless to say, again, all this is outside the scope of this book on relativity.

[65] Bohr, in Schilpp (ed.) [10], p. 235.

[66] For example, if you search the phrase, "in quantum mechanics, we are not dealing with an arbi-trary renunciation of a more detailed analysis of atomic phenomena," on Google Books, you immediately find well over a hundred books quoting Bohr.

panied Bohr on the trip, reports that there was little conversation between Bohr and Einstein during the visit. Indeed, when they did converse, "their conversation did not go beyond banalities." Apparently Einstein made it clear that the quantum topic was off-limits, and thus "Bohr was profoundly unhappy."[67] The so-called Bohr-Einstein debate was essentially over – at least directly between them.

By the late 1930s Bohr had advocated the use of "the word phenomenon [his emphasis] exclusively to refer to the observations obtained under specified circumstances, including an account of the whole experimental arrangement."[68] This was an extreme and clear exposition of an instrumentalist version of phenomenalism, with its emphasis on descriptions of the experimental conditions and even apparatuses employed in obtaining our knowledge of the world. To Einstein, this meant that our knowledge of that world was not direct, but limited by the medium (or means) of knowing – not just in an epistemological sense, but ontologically. For Einstein, on the contrary, a phenomenon must be described independently of the experimental apparatus. If quantum mechanics embraces such an ontological position, it must be incomplete in describing the real world.

Moreover, Einstein, with his grounding in philosophical literature, could see where this was going. In 1950, in another letter to Max Born (not Bohr), Einstein put his case this way: "you are after all convinced that no (complete) laws exist for a complete description [of reality, because of the statistical limitation of that knowledge], according to the positivistic maxim esse est percipi. Well, this is a programmatic attitude, not knowledge. This is where our attitudes really differ."[69] Einstein here clearly placed the issue within the realism/positivism framework, or what he called attitude.[70] Moreover, the maxim in Latin that he quoted is a famous line from the eighteenth-century Irish philosopher, George Berkeley, which means, "to be is to be perceived." In fact, there were those among the followers of Bohr (not Born) who went so far as to say that electrons and photons do not have an actual existence until they are measured – esse est percipi, indeed.[71]

---

[67] Quoted in Fölsing [65], p. 705.

[68] Bohr, in Schilpp (ed.) [10], pp. 237–238.

[69] Letter to Born, 15 September, 1950, in Einstein [56] [1950], p. 185.

[70] In a letter to Born (March 18, 1948), Einstein wrote that he wished to meet with him again, because, he wrote: "I would enjoy picking your positivistic philosophical attitude to pieces myself." Einstein [56] [1948], p. 160. It is true that Max Born made essential contributions the statistical interpretation of quantum physics. But that he held to the positivistic framework that Einstein accuses him of is debatable. Let me quote from a series of lectures Born delivered in 1948 (the same year as the above letter) where he discussed this very matter. Pointing out that "the question of reality cannot be avoided" in quantum physics, he confessed that he believed in "an external world which exists independently of us." Almost pleading his case he used the phrase "let me cling" to this idea. His argument began by positing that physics is fundamentally a search for invariants, and used the example of the charge and mass of an electron at rest. These "invariants of observation" led him to "maintain that the particles are real," and independent of our observation – "just a real," he wrote, as "a grain of sand." Born [13] [1948], pp. 103–105. I find these remarks exceedingly interesting.

[71] Brush [17], p. 420, notes that "it has taken some time for physicists and philosophers to realize that the position Einstein was defending was not merely classical determinism but, more significantly, common-sense realism. Most of us still find it hard to believe that the world has no real existence apart from ourselves...." Agreed!

Whether Bohr actually held such a radical view I do not know, since in much of his writings he talked around topics rather than facing them directly – thus making it difficult for the reader get beyond his rhetorical groping for answers. But surely such a conclusion is not too far a stretch from this sweeping statement attributed him: "There is no quantum world. There is only an abstract quantum physical description. It is wrong to think that the task of physics is to find out how nature is. Physics concerns what we can say about nature."[72] If nothing else, this is a succinct confirmation of my argument that the Bohr-Einstein debate may be understood in terms of epistemology and ontology, exposing Einstein's realism and the phenomenalism/positivism/instrumentalism of Bohr.[73]

* * *

This emphasis (or over emphasis) on philosophical aspects of interpreting quantum physics may mislead the reader into thinking this is the essence of its history. It is not. The history of the subject from 1900 into the 1930s is a triumph of many very successful experimental applications of the quantum hypothesis; its success in terms of empirical predictions and confirmations was dazzling, even to Einstein (as quoted above), who also made major contributions to the story. He, moreover, never said quantum physics was wrong, only that it was incomplete.

Infeld recalled that once when he pointed-out to Einstein that he had started the quantum theory, and thus asked him why he was so "dissatisfied" with it, Einstein replied with the quip, "Yes, I may have started it but I always regarded these ideas as temporary. I never thought that others would take them so much more seriously than I did."[74] This remark may be taken as playful, although Einstein did call his 1905 quantum idea of light "heuristic."[75]

Einstein's very close friend, Ehrenfest, however, was endlessly serious about the matter of the quantum, and especially to Einstein's realist resistance to Bohr's positivist interpretation that increasing set Einstein apart from mainstream physicists who were siding with Bohr. As a result Ehrenfest found himself in the middle of this

---

[72]Quoted in Petersen [166], p. 12. I should point out that this is not necessarily a direct quotation. Petersen was one of Bohr's assistants, and this quotation, rather like Bohr's epistemology, is Bohr's idea filtered through Petersen. Also quoted in Pais [163], pp. 426–427.

[73]As noted, I am avoiding details of quantum physics, by focusing on epistemology/ontology. But I do wish to note in passing the often deemed important paper published in 1935 that remains a source of much debate over the completeness of quantum physics. It was a collaboration among Einstein, Boris Podolsky, and Nathan Rosen. Einstein, et al. [58]. The argument is often referred to as the EPR paradox. There is extensive literature on this paper. In particular, the contemporary problem of what is called entanglement arose out of this paper; namely, the apparent ability of two subatomic particles to know where each one is forever after having interacted. I note in passing that Brush [17], p. 419 claims that Einstein, in fact, had little input to the paper (Einstein said Podolsky wrote most of it), and was less than satisfied with the argument. We do know that Einstein did not hold to the idea of entanglement, which he viewed as an extreme form of action-at-a-distance. In short, super-spooky.

[74]Quoted in Infeld [107], p. 110.

[75]Note the title: "On a Heuristic Point of View Concerning the Production and Transformation of Light." Recall, also, in his letter to Habicht at the time, that he called the idea "revolutionary."

debate. One physicist reports that around 1927 Ehrenfest came to him in tears because he had to make a choice between Einstein and Bohr, and that he agreed with Bohr.[76] To Ehrenfest, this was more than an intellectual debate in the scientific clouds; it was both a serious matter at the core of physics – and, as a physicist, at the core of his life – and an interpersonal problem at the heart of his relationships with both Bohr and Einstein.[77] It pained him, both to see Einstein become a scientific pariah, and to contribute to that isolation.[78]

Einstein, however, in some ways reveled in the isolation. To his old friend Besso he wrote in 1949: "I have become an obstinate heretic in the eyes of my colleagues."[79] That he was not troubled by this status may be inferred by what he wrote to Born around the same time. "I am generally regarded as a sort of petrified object, rendered blind and deaf by the years. I find this role not too distasteful, as it corresponds fairly well with my temperament."[80] To quote again Einstein's own personification, "God created the donkey and gave him a thick hide." To be sure, all contrarians need to be thick-skinned.

---

[76]Cited in Pais [162], p. 443.

[77]See Jones [110], *passim*. Ehrenfest's personal role navigating between Einstein and Bohr is a sub-theme running throughout Jones's superbly readable book. It is the best writing on Ehrenfest I know of since the first volume of M. J. Klein's biography (Klein [116]).

[78]In quoting from Einstein's eulogy to Ehrenfest in Chap. 8, I spoke of his tragic death. What Einstein did not speak of was this: The Ehrenfests had four children, the last, Vassily, had Down's syndrome, which was so severe that he was institutionalized at several hospitals for most of his life, the last being in Amsterdam. Over time Vassily's condition became a deeper burden on the family, emotionally and financially. In addition, Ehrenfest was prone to periods of severe depression, especially from about May 1931. There are some unanswered questions surrounding what happened on September 25, 1933, but what seems to have transpired is heartrending: Ehrenfest went to the institution in Amsterdam carrying a gun; he shot Vassily and then turned the gun on himself. See Jones [110], p. 285 and 311 note 54.

[79]Quoted in Pais [162], p. 462. Letter to Besso dated August 8, 1949. The reason, however, was not only his seemingly fruitless quest at unification but his objection to Bohr's interpretation of quantum physics. As he wrote in 1948 to another old friend, Habicht: "I still work indefatigably at science but I have become an evil renegade who does not wish physics to be based on probabilities." Quoted in Clark [26], p. 738.

[80]Einstein [56], p. 178. Letter the Born dated April 12, 1949.

Chapter 29
Legacy: From Pariah to Posthumous Prophet

"The theory does not yet contain the conclusions of the quantum theory. It furnishes, however, clues to a natural development, from which we may anticipate further results in this direction."[1] This statement from the 1931 report by Einstein on the unified field theory was quoted before at the end of Chap. 26, and I called it a curious sentence. It is curious because the quest for a unified field theory, as described it so far, was an attempt to unite gravity and electromagnetism. Where, or how, did quantum theory enter the topic, especially since we have seen Einstein being involved in a quarrel over the hegemony of Bohr's phenomenalist interpretation?

At Oxford University in 1933 – during his interregnum between Berlin and Princeton – Einstein delivered the Spencer lecture, from which was quoted the antipostivist creed: "I hold it true that pure thought can grasp reality, as the ancients dreamed." Further into the lecture he pointed to what he saw as a potential historical parallel. From the nineteenth to the twentieth century classical mechanics gave way to relativistic mechanics; importantly, the latter did not replace the former but rather extended it further into the realms of things traveling near the speed of light, where the old mechanics did not apply. In retrospect, therefore, the old mechanics was incomplete and required a modification by relativity. For example, we saw how relativistic effects reduce to classical physics when the speed of an object is much slower than light-speed.[2]

Quantum mechanics was incomplete (at least Einstein thought so) and required a modification, which a new theory would impart. But – and here is the crucial point he made – even though relativity in the end produced a modified version of Newtonian physics, it did not emerge from classical physics; that is, Einstein did not begin with the old physics and modify it to produce relativity, although initially he did try such tactics. Instead, as this book has shown, he went back to basics, setting an entirely new foundation, with a new set of postulates. Only through this new path was it possible to derive the relativistic modifications of classical physics. In view of that – and assuming an historical parallel – Einstein believed that a similar process

---

[1] Einstein [41], p. 439.

[2] Mathematically speaking, when $\mathbf{Q} = 1$.

D.R. Topper, *How Einstein Created Relativity out of Physics and Astronomy*, Astrophysics and Space Science Library 394, DOI 10.1007/978-1-4614-4782-5_29,
© Springer Science+Business Media New York 2013

of setting an entirely new foundation was necessary for the eventual modifications of quantum physics.

But what was the new foundation from which quantum mechanics would emerge as a by-product, so to speak? Well, a likely candidate was Einstein's own unified field theory, since it was based on new principles. Just as Newton's laws were integral to relativity, so Einstein hoped that quantum laws would emerge out of the final unified theory. Such a theory would then be fully unified: gravity, electromagnetism, and the quanta – all in one. Thus the sentence above from the 1931 report is even more than curious: it is expressing a dream, a vision that never came true during his life, despite his quest to the very end. For that reason it is worth quoting again: "The [potential unified] theory does not yet contain the conclusions of the quantum theory. It furnishes, however, clues to a natural development, from which we may anticipate further results in this direction." That was his hope: not only to unify gravity and electromagnetism, but by additionally deducing quantum physics, to, in turn, complete that incomplete theory.[3]

Einstein's unfulfilled fantasy made him a pariah during most of his Princeton years, only ceasing with his death in 1955. Yet the quest did not die with him. It was revived in the early 1970s after the unification of the three forces within the atom, at first predicted and then experimentally verified. With the unification of the strong nuclear, the weak nuclear, and the electrical forces, only gravity was left hanging. Not surprisingly, some gutsy physicists began probing the possibility of bringing gravity into the mix – and so Einstein's last quest was reborn about twenty years after he died. How ironic it is that out of nuclear physics – a subject Einstein ignored his entire life – the unification quest for all forces of nature was reborn.[4]

Einstein, intriguingly, predicted this revival in a letter to Solovine in 1948, when he confessed that he "shall never solve" the unified theory; but he added: "it will fall into oblivion and be discovered anew later."[5] The most famous and well known physicist pursuing this today is, certainly, Stephen Hawking, who, as I write this, has not achieved that goal.[6] Neither have myriad others, some postulating superstrings, others

---

[3] A commonly use metaphor for Einstein's thirty-year pursuit has been to refer to it as his unfinished symphony. The comparison is surely with Schubert's Eighth symphony and as such is, to me, a very poor analog. It was not Schubert's last symphony, and it was not unfinished due to this death. A better analog would be Johan Sebastian Bach's last fugue, which is seldom played since it stops in the middle of a musical phrase. I discuss this in more detail in Topper [198], p. 152.

[4] Holton [100], p. 166.

[5] Einstein [54], p. 107. Letter of November 25, 1948.

[6] Steven Weinberg (one of the unifiers of the electric, weak and strong nuclear forces, and who shared in the Nobel Prize) reviewed a recent book by Hawking, *The Grand Design*, written with Leonard Mlodinow. In his review, Weinberg points to Hawking's "disturbing" idea there may be a number of equally valid theories of reality, and hence there is no real "underlying theory." Weinberg admits that "the nature of reality" has "puzzled scientists and philosopher for millennia." His own position echoes Einstein, as he writes: "I think that there is something real out there, entirely independent of us and our models.... But this is because I can't help believing in an objective reality, not because I have good arguments for it. I am in no position to argue that Hawking's antirealism [or positivism] is wrong. But I do insist that neither quantum mechanics nor anything else in physics settles the question." Weinberg [207], p. 32.

conceiving yet unknown subatomic particles that may hold the key to unifying all four forces of nature. I like the way Canadian physicist, Lee Smolin put it: the Princeton Institute where Einstein was a pariah late in life "is now filled with theorists who search for new variants of unified field theories. It is indeed a vindication of sorts for Einstein...."[7] This neo-quest is called the search for a theory of everything or alternately the holy grail of physics.[8] Tying-up relativity and quantum mechanics into one package is no longer a sideshow of physics. Some of the best minds in the science are diligently pursuing this, as the quest moves into the mainstream. Einstein was seen as a pariah in his late years, but in retrospect he was probably a prophet – especially if one those "best minds" eventually finds THE theory.

Einstein surely had one of the best minds of all time. Nonetheless, he did not see it that way. He was probably at least partially serious when he said:

> I am no more gifted than anybody else. I am just more curious than the average person and I will not give up on a problem until I have found the proper solution. This is one of my greatest satisfactions in life – solving problems – and the harder they are, the more satisfaction do I get out of them. [9]

His ultimate goal – as for Kepler and Newton – was to probe the mind of what he alternately called God or "the Old One," whatever he meant by that.

My goal here, needless to say, was much less ambitious – although for me, perhaps almost as daunting – namely, to probe Einstein's mind, and to explain how he discovered relativity in physics and astronomy from Galileo to Hubble, with the goal of understanding and therefore appreciating what he had wrought.[10]

* * *

Those who worked with Einstein in his later years at the Institute on the unification quest often mentioned a little quirky behavior of the man. When they were stuck in their work, with no clear direction where to move in a theoretical argument or equation, Einstein would stand still or usually pace back and forth, twirl a lock of hair, and say in his broken English, "I will a little think" – which more often came out as

---

[7] Smolin [187], p. 40.

[8] See Greene [82], p. 15, who concedes: "Einstein was simply ahead of his time.... [H]is dream of a unified theory has become the Holy Grail of modern physics." Greene, along with many writers on Einstein's quest, also refers to the unity search as a "quixotic quest," which is a comparative reference to the fictional character of Don Quixote. Sayen [178], p. 134 even says that Einstein identified with the fictional knight. Recall too that Solovine in the introduction to his letters (Einstein [54], p. 9) reports that the Olympia Academy read Cervantes' book; and Infeld [106], pp. 312–313 says that it was Einstein's favorite book of fiction, and that a copy was on his night table "for relaxation." Personally I abhor the analogy. I find the stupidity of Don Quixote at odds with my image of Einstein, and hence it is difficult for me to conceive of Einstein identifying with the foolish knight. Accordingly, you will neither find me calling Einstein's quest quixotic nor an unfinished symphony (see footnote 3 above).

[9] Quoted in Bucky [22], p. 29.

[10] McCormmach [138], is still a very brief and very valuable overview of the many historiographical ways of approaching Einstein's life and work.

"I will a little tink." In time, he would stop, smile, look at them, and he usually had an answer.[11]

Another idiosyncrasy of his was that he occasionally replied to correspondents with a short doggerel verse. As an example, here is a poem he sent from California in 1933 to the Belgium Queen who, recall,[12] he was friends with:

> In cloister garden a small tree stands.
> Planted by your very hands.
> It sends – its greetings to convey –
> A twig, for it itself must stay.[13]

In the tradition of Edmund Halley – he of Halley's Comet fame – who wrote a poem to Newton at the commencement of the Principia[14]: I submit, in homage to the hero of my book – albeit, a patently flawed one[15] – and as a finale, my feeble attempt at a doggerel verse.

> This author, most undoubtedly, would be tickled pink,
> If via pixels or, more likely, plain ol' paper & ink,
> This book spurred, in an occasional reader, "a little tink"
> On relativity theory – by and large simple, yet sublime,
> And the struggles of its maker – my somewhat
> tarnished hero, Einstein.

---

[11] I have read variations of this behavior in a number of sources over the years. For one, see Hoffmann, in Woolf [215], pp. 477–478.

[12] Recall too that later in the same year Einstein and his family would be in Belgium, being guarded from Nazi assassins.

[13] Quoted in Dukas and Hoffmann (eds.) [35], pp. 48–49. The original German (p. 135) is: *Ein Baum im Klostergarten stand/Der war gepflanzt von Ihrer Hand./Ein Zweiglein sendet er zum Gruss,/Weil er dort stehen bleiben muss.*

[14] Halley had every right to write whatever he wished, since in essence he paid for the publication of the book. For details see Topper [198], pp. 155–158.

[15] As a final comment in this last footnote: I make no excuses for this unabashedly old-fashioned heroic (scientific) biography of Einstein – the pervasiveness of postmodern historiography not withstanding.

Bibliography

1. Abiko, Seiya. 2000. Einstein's Kyoto address: 'how I created the theory of relativity'. *Historical Studies in the Physical Sciences* 31(pt.1): 1–35.
2. Adler, Carl G. 1987. Does mass really depend on velocity, dad? *American Journal of Physics* 55(8): 739–743.
3. Alpher, Ralph A. 1998. Message on the Internet to the History of Astronomy Discussion Group on Einstein's "blunder" over the cosmological constant, 2 Apr 1998.
4. Arnheim, Rudolf. 1969. *Visual thinking*. London: Farber and Farber.
5. Baigrie, Brian. 2007. *Electricity and magnetism: a historical perspective*. Westport: Greenwood Press.
6. Barbour, Julian, and Herbert Pfister (eds.). 1995. *Mach's principle: from Newton's bucket to quantum gravity*, Einstein Studies, vol. 6. Boston/Basel/Berlin: Birkhäuser.
7. Barnett, Lincoln. 1948. *The universe and Dr. Einstein*. New York: Mentor Books.
8. Bartusiak, Marcia. 2009. *The day we found the universe*. New York: Pantheon Books.
9. Bernstein, Jeremy. 1996. *A theory for everything*. New York: Copernicus (Springer-Verlag).
10. Bohr, Niels. 1949. Discussion with Einstein on epistemological problems in atomic physics. In P.A. Schilpp (ed.), 1949, vol. I, 199–241.
11. Bondi, Hermann, and Joseph Samuel. 1996. The lense–thirring effect and Mach's principle. Available online at: arXiv:gr-qc/96070-09v1 4 July 1996.
12. Born, Max. 1949. Einstein's statistical theories. In P.A. Schilpp (ed.), 1949, vol. I, 161–177.
13. Born, Max. 1964. *Natural philosophy of cause and chance*. New York: Dover. This is reprint of a series of lectures delivered at Oxford University in 1948.
14. Born, Max. 2005. *The Born-Einstein Letters*, see Einstein 2005, below.
15. Brian, Denis. 1996. *Einstein: a life*. New York: Wiley.
16. Brush, Stephen. 1979. Einstein and indeterminism. *Journal of the Washington Academy of Science* 69(3): 89–94.
17. Brush, Stephen. 1980. The chimerical cat: philosophy of quantum mechanics in historical perspective. *Social Studies of Science* 10: 393–447.
18. Brush, Stephen. 1999. Why was relativity accepted? *Physics in Perspective* 1(2): 184–214.
19. Brush, Stephen. 2002. Cautious revolutionaries: Maxwell, Planck, Hubble. *American Journal of Physics* 70(2): 119–127.
20. Brush, Stephen. 2003. Review of the Cambridge history of science. Vol. 5. *The modern physical and mathematical sciences,* ed. Mary Jo Nye, Cambridge: Cambridge University Press, in Isis, vol. 94, no. 4, 687–688.
21. Brush, Stephen. 2007. How ideas became knowledge: the light-quantum hypothesis 1905–1935. *Historical Studies in the Physical and Biological Sciences* 37(2): 205–246.
22. Bucky, Peter A. 1992. *The private Albert Einstein*. Kansas City: Andrews and McMeel.

D.R. Topper, *How Einstein Created Relativity out of Physics and Astronomy*, Astrophysics and Space Science Library 394, DOI 10.1007/978-1-4614-4782-5,
© Springer Science+Business Media New York 2013

23. Calder, Nigel. 1983. *Einstein's universe: a guide to the theory of relativity.* New York: Penguin Books.
24. Canales, Jimena. 2005. Einstein, Bergson, and the experiment that failed: intellectual cooperation at the League of Nations. *Modern Language Notes* 120: 1168–1191.
25. Cassidy, David C. 2004. *Einstein and our world*, 2nd ed. New York: Humanity Books.
26. Clark, Ronald W. 1971. *Einstein: the life and times.* New York: Avon Books.
27. Cohen, I. Bernard. 1955. An interview with Einstein. *Scientific American* 193(1): 68–73 (July). This interview was conducted two weeks before Einstein died.
28. Crelinsten, Jeffrey. 1980. Einstein, relativity, and the press: the myth of incomprehensibility. *The Physics Teacher* 18(February): 115–122; Physicists receive relativity: revolution and reaction. *The Physics Teacher* 18(March): 187–193. This is a two-part article on how the idea that relativity was an incomprehensible theory arose in the popular press and among some physicists themselves.
29. Crelinsten, Jeffrey. 1983. William Wallace Campbell and the 'Einstein Problem': an observational astronomer confronts the theory of relativity. *Historical Studies in the Physical Sciences* 14(pt. 1): 1–91.
30. Crelinsten, Jeffrey. 2006. *Einstein's jury: the race to test relativity.* Princeton: Princeton University Press.
31. Darnton, Robert. 1968. *Mesmerism and the end of the enlightenment in France.* Cambridge, MA: Harvard University Press.
32. Darrigol, Olivier. 2003. Quantum theory and atomic structure, 1900–1927. The Cambridge history of science. Vol. 5, *The modern physical and mathematical sciences,* ed. Mary Jo Nye. Cambridge: Cambridge University Press. Darrigol's essay is Chapter 17.
33. Dewhirst, David, and Michael Hoskin. 1997. The message of starlight: the rise of astrophysics. In *The Cambridge illustrated history of astronomy,* ed. Michael Hoskin, 256–343. Cambridge: Cambridge University Press.
34. Duerbeck, Hilmar W., and Waltraut C. Seitter. 2001. In Hubble's shadow: early research on the expansion of the universe. In *Miklós Konkoly Thege (1842–1916). 100 years of observational astronomy and astrophysics: a collection of papers on the history of observational astrophysics,* ed. C. Sterken and J.B. Hearnshaw, 231–254. Brussels: VUB Press. This is a rather obscure publication; fortunately it available on-line: http://homepages.vub.ac.be/~hduerbec/hubbleshadow.pdf.
35. Dukas, Helen, and Banesh Hoffmann (eds.). 1979. *Albert Einstein: the human side: new glimpses from his archives.* Princeton: Princeton University Press.
36. Dyson, Freeman. 2003. Clockwork science. *New York review of books* (November 6), 42–44. This is an essay review of Peter Galison's *Einstein's Clocks, Poincaré's Maps*, below.
37. Eddington, Arthur S. 1963. *The mathematical theory of relativity.* Cambridge: Cambridge University Press. This is a reprint of the second edition of 1924. The first edition appeared in 1923.
38. Einstein, Albert. 1923. Fundamental ideas and problems of the theory of relativity. In *Noble lectures: physics, 1901–1921.* Vol. I. New York: Elsevier, 1967, 483–490.
39. Einstein, Albert. 1930. Raum, Äther und Feld in der Phyisk ("Space, Aether, and Field in Physics"). *Forum Philosophicum* 1: 173–184. An English translation by Edgar S. Brightman, 180–184.
40. Einstein, Albert. 1931a. Professor Einstein at the California Institute of Technology: addresses at the dinner in his honor. *Science* 73(1893): 375–381 (April 10).
41. Einstein, Albert. 1931b. Report: gravitational and electromagnetic fields. *Science* 74(1922): 438–439 (October).
42. Einstein, Albert. 1934. *Essays in science,* Trans. Alan Harris. New York: Philosophical Library. This is an abridged English translation of the Mein Weltbild (1934), consisting mainly of scientific essays. See Einstein, 1949b, below. Also see Einstein, 1954.
43. Einstein, Albert. 1949a. Dr. Albert Einstein and American colleagues, 1931. *Proceedings of the American Philosophical Society,* vol. 93, no. 7 (December), 543–545. This is a copy of

Einstein's handwritten lecture (in German), with an English translation, which he delivered at the Caltech faculty club, the Athenaeum, on January 15, 1931.

44. Einstein, Albert. 1949b. *The world as I see it.* Trans. Alan Harris. New York: The Wisdom Library. This is a collection of essays from 1922 to 1934. It is an abridged English translation of the earlier published Mein Weltbild (1934), leaving-out the scientific writings. See the second English (companion) anthology *Out of My Later Years*, 1950b. See also 1934, above, and 1954, below.

45. Einstein, Albert. 1950a. On the generalized theory of gravitation. *Scientific American* vol. 183 (April), 258–262.

46. Einstein, Albert. 1950b. *Out of my later years.* No translator cited. New York: The Wisdom Library. This is a collection of essays from 1934 to 1950. It is a second (companion) anthology to *The World as I See It*, 1949b.

47. Einstein, Albert. 1954. *Ideas and opinions.* Trans. and revisions by Sonja Bargmann. New York: Bonanza Books. Based, in part, on Mein Weltbild. Edited by Carl Seelig, and others (Amsterdam: Querido Verlag, 1934); plus a further edition by Seelig published in Switzerland in 1953, and other sources, such as *Out of My Later Years* (1950), cited above. Seelig's 1934 German edition was translated by Alan Harris as *The World As I See It* (but recent editions, see above, leave-out the scientific essays). See 1934, above, for translations of some of the scientific articles. Many of the essays in *Ideas and Opinions* do not cite original sources. According to Schilpp (ed.), 1949, vol. II, p.737, Seelig "gives no clue as to where items were originally published; some may never have appeared in print previously."

48. Einstein, Albert. 1956. *The meaning of relativity,* 5th ed. Princeton University Press. This book contains the four lectures delivered at Princeton University in 1921 (pp. 1–108). Translated by Edwin P. Adams. An Appendix was added to the second edition of 1945 (pp. 109–132). Translated by Ernst G. Straus. For the third edition in 1950 another Appendix II was added, which was revised for the fourth edition (1953), and for the fifth edition Einstein completely revised this Appendix in December 1954, about four months before he died. This last version of Appendix II he titled: "Relativistic Theory of the Non-Symmetric Field" (pp. 133–166). Translated by Sonja Bargmann. This last paper was written with Bruria Kaufman, an Israeli physicist and his last collaborator.

49. Einstein, Albert. 1960. *Relativity: the special and the general theory,* 15th ed. Trans. Robert W. Lawson in 1920. London: Methuen & Co. This popular account was first published in German in 1917. This edition has five appendices, the last (1952) is titled "Relativity and the Problem of Space."

50. Einstein, Albert. 1968. *Einstein on peace,* ed. Otto Nathan and Heinz Norden. New York: Schocken Books. Also cited below, Nathan, 1968.

51. Einstein, Albert. 1979. *Autobiographical notes.* Trans. and ed. Paul A. Schilpp. La Salle & Chicago: Open Court Publishing. This is the corrected version of the original 1947 German manuscript, first published in 1949. The uncorrected version is the more accessible one: see *Albert Einstein: Philosopher-Scientist.* Two volumes. Edited by Paul A. Schilpp. New York: Harper & Row, 1949, vol. I, 3–95. The latter book is cited separately below under Schilpp (ed.), 1979.

52. Einstein, Albert. 1982. How I created the theory of relativity. *Physics Today* 35: 45–47 (August). This is a transcription by Jun Ishiwara of a talk Einstein gave in Kyoto University on December 14, 1922, and translated by Yoshimasa A. Ono. A corrected translation is in Abiko, 2000. Skepticism on the validity of this document is in Miller, 1987 and Holton, 1988.

53. Einstein, Albert. 1983. Aether and the theory of relativity, *Sidelights on relativity.* New York: Dover Publications, 3–24. This is a lecture delivered on October 27, 1920 at the University of Leiden. It is a reprint of a 1922 translation by G. B. Jeffrey and W. Perrett. Another translation of this essay appears in Einstein, 1934, above, as "Relativity and the Ether," 98–111.

54. Einstein, Albert. 1986. *Letters to Solovine. Introduction by Maurice Solovine.* Trans. Wade Baskin. New York: Philosophical Library. This useful volume contains almost all the original German photocopies (or otherwise German transcripts) along with the English translations.

55. Einstein, Albert. 1987 +. *The collected papers of Albert Einstein*. Princeton: Princeton University Press. As I write this, the series has been published through vol. 12 (to 1921). Each document is cited in the text as: Einstein Papers, Vol. #, Doc. #.

56. Einstein, Albert. 2005. *The Born-Einstein letters: friendship, politics and physics in uncertain times*. Trans. Irene Born. New York: Macmillan. This is a collection of correspondence between Einstein and Max and Hedwig Born from 1916 to 1955, with commentaries by Max Born.

57. Einstein, Albert, H.A. Lorentz, H. Weyl, and H. Minkowski. 1923. *The principle of relativity*. New York: Dover Publications. This book contains the first English transitions of essential papers on relativity. The translations of Einstein's papers in *The Collected Papers of Albert Einstein* have supplanted some of these; in other cases the *Einstein Papers* merely reprint these translations of 1923.

58. Einstein, Albert, B. Podolsky, and N. Rosen. 1935. Can quantum-mechanical description of physical reality be considered complete. *Physical Review* 47: 777–780. The paper is reprinted, along with Bohr's rebuttal, in *Physical reality: philosophical essays on twentieth-century physics*. Edited by Stephen Toulmin. New York: Harper Torchbooks, 1970, 122–130. Bohr's essay is on pp. 130–142. Bohr's paper, with the identical title, was also published in the *Physical Review* 48 (1935): 696–702.

59. Einstein, Albert, and Leopold Infeld. 1961. *The evolution of physics: the growth of ideas from early concepts to relativity and quanta*. New York: Simon & Schuster. This was first published in 1938. In the preface to the 1961 edition, Infeld acknowledges Einstein as the "chief author" of the book.

60. Eisenstaedt, Jean. 2006. *The curious history of relativity: how Einstein's theory of gravity was lost and found again*. Trans. Arturo Sangalli. Princeton/Oxford: Princeton University Press.

61. Farrell, John. 2005. *The day without yesterday: Lemaître, Einstein, and the birth of modern cosmology*. New York: Thunder's Mouth Press.

62. Feuer, Lewis S. 1982. *Einstein and the generations of science*, 2nd ed. New Brunswick: Transaction Books.

63. Feynman, Richard P., Robert B. Leighton, and Matthew Sands. 1963. *The Feynman lectures on physics*. Reading: Addison-Wesley.

64. Fishbane, Paul. 2010. Time warp: recent right-wing rejections of Einstein's theory of relativity Echo Nazi Dismissals of What They called 'Jewish Physics'. *Tablet Magazine*. This article was downloaded 12 Nov 2010 at: http://www.tabletmag.com/news-and-politics/50097/time-warp/.

65. Fölsing, Albrecht. 1997. *Albert Einstein: a biography*. Trans. Ewald Osers. New York: Viking. Originally published in German in 1993.

66. Fowles, Grant R. 1962. *Analytical mechanics*. New York: Holt, Rinehart, and Winston.

67. Frank, Philipp. 1947. *Einstein: his life and times*. Trans. George Rosen. Edited and revised by Shuichi Kusaka. New York: Alfred A. Knopf.

68. French, A.P. (ed.). 1979. *Einstein: a centenary volume*. Cambridge: Harvard University Press.

69. Friedman, Alan J., and Carol C. Donley. 1985. *Einstein as myth and muse*. Cambridge: Cambridge University Press.

70. Galilei, Galileo. 1957. Letters on sunspots. *Discoveries and Opinions of Galileo*, pp. 87–144. Ed. and trans. Stillman Drake. New York: Doubleday. The letters were published in 1613.

71. Galilei, Galileo. 1967. *Dialogue concerning the two chief world systems*. Trans. Stillman Drake. Berkeley: University of California Press. The Dialogue was published in 1632.

72. Galilei, Galileo. 1989. *Sidereus Nuncius or The Sidereal Messenger*. Trans. Albert Van Helden. Chicago: University of Chicago Press. The Sidereus Nuncius was published in 1610.

73. Galison, Peter L. 1979. Minkowski's space-time: from visual thinking to the absolute world. *Historical Studies in the Physical Sciences* 10: 85–121.

74. Galison, Peter L. 2000. Einstein's clocks: the place of time. *Critical Inquiry* 26(2): 355–389 (winter).
75. Galison, Peter L. 2003. *Einstein's clocks, poincaré's maps: empires of time.* New York: W. W. Norton.
76. Galison, Peter L. 2004. Einstein's compass. *Scientific American* 291(3): 66–69 (September).
77. Peter L. Galison. 2008. The Assassin of relativity. *Einstein for the 21st century: his legacy in science, art, and modern culture,* ed. Peter L. Galison, Gerald Holton, and Silvan S. Schweber. Princeton: Princeton University Press, 185–204. This essay is on Einstein's friendship with Friedrich Adler.
78. Gamow, George. 1970. *My world line: an informal autobiography.* New York: Viking Press.
79. Grafton, Anthony. 1997. *The footnote: a curious history.* Cambridge, MA: Harvard University Press.
80. Graney, Christopher M. 2008. But still, it moves: tides, Stellar parallax, and Galileo's commitment to the Copernican theory. *Physics in Perspective* 10: 258–268.
81. Green, Jim (ed.). 2003. *Albert Einstein.* Melbourne/New York: Ocean Press. From the series, Rebel Lives, this book consists of a short introduction, followed by excerpts from his writings on mainly social and political topics.
82. Greene, Brian. 2003. *The elegant universe: superstrings, hidden dimensions, and the quest for the ultimate theory.* New York: W. W. Norton & Company.
83. Grundmann, Siegfried (ed.) 2005. *The Einstein dossiers: science and politics – Einstein's Berlin period with an appendix on Einstein's FBI file.* Trans. Ann M. Hentschel. New York/ Berlin: Springer.
84. Gunter, P.A.Y. (ed. and trans.). 1969. *Bergson: and the evolution of physics.* Knoxville: University of Tennessee Press.
85. Hadamard, Jacques. 1954. *The psychology of invention in the mathematical field.* New York: Dover Publications. This is a reprint of the book first published in 1945, and enlarged in 1949.
86. Halpern, Paul. 2007. Klein, Einstein, and five dimensional unification. *Physics in Perspective* 9: 390–405.
87. Harman, P.M. 1982. *Energy, force, and matter: the conceptual development of nineteenth-century physics.* Cambridge: Cambridge University Press.
88. Hentschel, Klaus. 1992. Einstein's attitude towards experiments: testing relativity theory, 1907–1927. *Studies in History and Philosophy of Science* 23(4): 593–624.
89. Hentschel, Klaus. 1993. The conversion of St. John: a case study on the interplay of theory and experiment. *Einstein in context,* ed. Mara Beller, Robert S. Cohen, and Jürgen Renn. Cambridge: Cambridge University Press, 137–194.
90. Herbert, Christopher. 2001. *Victorian relativity: radical thought and scientific discovery.* Chicago: University of Chicago Press.
91. Hertz, Heinrich. 1962. *Electric waves: being researches on the propagation of electric action with finite velocity through space.* Trans. D.E. Jones. New York: Dover Publications. This is a reprint of the original edition of 1893.
92. Hetherington, Norriss S. 1988. Sirius B and the gravitational redshift. In *Science and objectivity: episodes in the history of astronomy.* Ames: Iowa State University Press, Chapter 6, 65–72.
93. Highfield, Roger. 2005. Einstein's women. In Renn (ed.), 2005, below, vol. I, 242–263.
94. Highfield, Roger, and Paul, Carter. 1993. *The private lives of Albert Einstein.* London/Boston: Faber & Faber. Written by two journals, this is certainly the most muckraking account of Einstein's personal life. Although they often take gossip as fact, much of their interpretation of real documentation is reasonable and persuasive.
95. Hoefer, Carl. 1994. Einstein's struggle for a Machian gravitation theory. *Studies in History and Philosophy of Science* 25(3): 287–335.
96. Hoefer, Carl. 1995. Einstein's formulation of Mach's principle. In ed. J. Barbour and H. Pfister, 1995, above, 67–90.

97. Hoffmann, Banesh. 1972. *Albert Einstein: creator and rebel*. New York: New American Library.

98. Holton, Gerald. 1981. Einstein's search for the Weltbild. *Proceedings of the American Philosophical Society* 125(1): 1–15 (February).

99. Holton, Gerald. 1988. *Thematic origins of scientific thought: Kepler to Einstein*. Revised edition. Cambridge, MA: Harvard University Press. Chapter 8 (pp. 279–370) is a reprint of the classic article, "Einstein, Michelson, and the 'Crucial' Experiment," published in Isis, vol. 60 (1969), 133–197. Comments on subsequent work are on 477–480.

100. Holton, Gerald. 1996. *Einstein, history, and other passions: the rebellion against science at the end of the twentieth century*. Reading: Addison-Wesley.

101. Holton, Gerald. 2005. The woman in Einstein's shadow. In ed. Renn, 2005, below, vol. I, 332–335.

102. Holton, Gerald, and Yehuda Elkana (eds). 1982. *Albert Einstein: historical and cultural perspectives: the centennial symposium in Jerusalem*. Princeton: Princeton University Press. The Symposium was held March 14–23, 1979.

103. Hoskin, Michael. 1997. The astronomy of the universe of stars. In ed. Michael Hoskin *The Cambridge illustrated history of astronomy*. Cambridge: Cambridge University Press, 198–255.

104. Howard, Don. 1993. Was Einstein really a realist? *Perspectives on Science* 1(2): 204–251.

105. Hubble, Edwin. 1953. The law of red-shifts. *Monthly Notices of the Royal Astronomical Society* 113: 658–666. This is the George Darwin Lecture, delivered on May 8, 1953.

106. Infeld, Leopold. 1941. *Quest: the evolution of a scientist*. New York: Doubleday, Doran, & Co.

107. Infeld, Leopold. 1955. *Albert Einstein: his work and influence on our world*. Revised edition. New York: Scribner's Sons. The first edition was published in 1950. The revised edition is not dated, but its starts by mentioning Einstein's death in 1955. Hence I have chosen that date.

108. Infeld, Leopold. 1961. See Einstein and Infeld.

109. Isaacson, Walter. 2007. *Einstein: his life and universe*. New York: Simon & Schuster.

110. Jones, Sheila. 2008. *The quantum ten: a story of passion, tragedy, ambition, and science*. Toronto: Thomas Allen.

111. Jungnickel, Christa, and Russell McCormmach. 1986. *Intellectual mastery of nature: theoretical physics from Ohm to Einstein, vol. II, the now mighty theoretical physics, 1870–1925*. Chicago/London: University of Chicago Press.

112. Kahn, Carla, and Franz Kahn. 1975. Letters from Einstein to de Sitter on the nature of the universe. *Nature* 257: 451–454 (October 9).

113. Kant, Immanuel. 1970. *Metaphysical foundations of natural science*. Trans. James Ellington, from *Die Metaphysischen Anfangsgründe der Naturwissenschaft*, 1786. Indianapolis/New York: Bobbs-Merrill.

114. Kayser, Rudolf. 1946. *Spinoza: portrait of a spiritual hero*. Introduction by Albert Einstein. Trans. Amy Allen and Maxim Newmark. New York: Philosophical Library. See Reiser, 1930.

115. Kerszberg, Pierre. 1989. *The invented universe: the Einstein-De Sitter controversy (1916–17) and the rise of relativistic cosmology*. Oxford: Clarendon Press.

116. Klein, Martin J. 1970a. *Paul Ehrenfest: vol. I: the making of a theoretical physicist*. New York/Amsterdam: North-Holland. Klein died before completing volume two. Sheilla Jones (Jones, 2008) told me that in speaking with Klein on why the second volume was never written, he implied that he was psychologically blocked because of having to confront Ehrenfest's suicide and murder of his son.

117. Klein, Martin J. 1970b. The first phase of the Bohr-Einstein dialogue. *Historical Studies in the Physical Sciences* 2: 1–39.

118. Klein, Martin J. 1975. Einstein on scientific revolutions. *Vistas in Astronomy* 17: 113–120.

119. Klein, Martin J., A. Shimony, and T.J. Pitch. 1979. Paradigm lost? A review symposium. *Isis* 70(253): 429–440; an essay review of Kuhn's 1978 edition of *Black-Body Theory and the Quantum Discontinuity, 1894–1912* (Chicago). See Kuhn, 1987, below.

120. Koyré, Alexander. 1957. *From the closed world to the infinite universe*. Baltimore/London: Johns Hopkins University Press.
121. Kragh, Helge L. 1996. *Cosmology and controversy: the historical development of two theories of the universe*. Princeton: Princeton University Press.
122. Kragh, Helge L. 2007. *Conceptions of cosmos: from myths to the accelerating universe: a history of cosmology*. New York/Oxford: Oxford University Press.
123. Kragh, Helge, and Robert W. Smith. 2003. Who discovered the expanding universe? *History of Science* 41(2): 141–162 (June).
124. Kuhn, Thomas S. 1962. *The structure of scientific revolutions*. Chicago: University of Chicago Press. See also the Second enlarged edition of 1970.
125. Kuhn, Thomas S. 1977. Energy conservation as an example of simultaneous discovery. In *The essential tension: selected studies in scientific tradition and change*, ed. T.S. Kuhn., 66–104 Chicago: University of Chicago Press. The article was originally published in 1959.
126. Kuhn, Thomas S. 1987. *Black-body theory and the quantum discontinuity, 1894–1912*. Chicago/London: University of Chicago Press. This is a reprint of the 1978 edition, with a new Afterword.
127. Leavitt, Henrietta S. 1908. 1777 variables in the magellanic clouds. *Annals of Harvard College Observatory* 60(4): 87–103.
128. Leavitt, Henrietta S. 1912. Periods of 25 variable stars in the small magellanic cloud. *Annals of Harvard College Observatory* 173(March): 1–3. In this article, the author is listed as Edward C. Pickering, who was the director of the observatory, and commonly used his name thusly in publications from the institution. Nonetheless, it begins by noted that the paper was "prepared by Miss Leavitt."
129. Lemaître, Georges. 1931. A homogeneous universe of constant mass and increasing radius accounting for the radial velocity of extra-galactic nebulae. *Monthly Notices of the Royal Astronomical Society* 91(March): 483–489. This is an English translation of the original article of 1927 published in French in Brussels, in the rather obscure journal, *Anneles Scientifique Bruxelles*. But the English translation left out some of the text and most of the footnotes: see Livio, 2011, below.
130. Lemaître, Georges. 1949. The cosmological constant. In Schilpp (ed.), 1949, vol. II, 437–456.
131. Lerner, Eric. J., and José B. Almeida (eds.). 2006. 1st crisis in cosmology conference, CCC-I. (June 23–25, 2005) *AIP conference proceedings*, vol. 822.
132. Levenson, Thomas. 2003. *Einstein in Berlin*. New York: Bantam Books.
133. Livio, Mario. 2011. Lost in translation: mystery of the missing text solved. *Nature* 479 (November 10): 171–173. On Lemaître's classic 1931 article; see Lemaître, 1931, above.
134. Mach, Ernst. 1960. *The science of mechanics: a critical and historical account of its development*, 6th ed. Trans. Thomas J. McCormack. New York: Open Court. The original German edition, Die Mechanik in Inrer Entwicklung, Historisch-Kritisch Dargestellet, was published in 1883. The German edition went through nine revisions.
135. Marianoff, Dimitri, with Palma Wayne. 1944. *Einstein: an intimate study of a great man*. New York: Doubleday, Doran, and Co. Marianoff was Einstein's son-in-law, married to Elsa's daughter, Margot, but later divorced.
136. Marić, Mileva. 2003. *In Albert's shadow: the life and letters of Mileva Marić*, ed. Milan Popović. Translated from the Serbian and German by Boško Milosavljević and Branimir Živojinović. Baltimore: The Johns Hopkins University Press.
137. Martinez, Alberto A. 2005. Handling evidence in history: the case of Einstein's wife. *School Science Review* 86(March): 49–56.
138. McCormmach, Russell. 1970. Editor's forward. *Historical studies in the physical sciences*, vol. 2, ix–xx. This very brief yet brilliant essay is still one of the best short introductions to Einstein's intellectual and professional life.
139. McFarland, Ernie. 1998. *Einstein's special relativity: discover it for yourself*. Toronto: Trifolium Books.
140. McVittie, George C. 1967. (Obituary notice of) George Lemaître. *Quarterly Journal of the Royal Astronomical Society* 8: 294–297.

141. Michelmore, Peter. 1962. *Einstein: profile of the man*. New York: Dodd, Mead & Co. Michelmore draws heavily on interviews with Einstein's son, Hans, as well as Helen Dukas and Otto Nathan, who were the trustees of Einstein's estate.

142. Miller, Arthur I. 1987. Einstein and Michelson-Morley. *Physics Today* 40: 8–13 (May).

143. Miller, Arthur I. 1998. *Albert Einstein's special theory of relativity: emergence (1905) and early interpretation (1905–1911)*. Republished edition; original 1981. New York: Springer.

144. Miller, Arthur I. 2000. *Insights of genius: imagery and creativity in science and art*. Cambridge, MA: MIT Press.

145. Millikan, Robert A. 1963. *The electron*. Chicago: University of Chicago Press. This is a reprint of the original 1917 edition.

146. Minkowski, Hermann. 1908. Space and time. Reprinted in Einstein et al. 1923, *The principle of relativity*, 73–96.

147. Moszkowski, Alexander. 1970. *Conversations with Einstein*. Trans. Henry L. Brose. New York: Horizon Press. Originally published in German in 1921. This book is based on a series of interviews by the author from the summer of 1919 to the autumn of 1920.

148. Nathan, Otto, and Heinz Norden (eds.). 1968. *Einstein on peace*. New York: Schocken Books.

149. Neffe, Jürgen. 2007. *Einstein: a biography*. Trans. Shelley Frisch from the 2005 German edition. New York: Farrar, Straus, & Giroux.

150. Newton, Isaac. 1969. *De Mundi systemate (A treatise on the system of the world)*. English translation. London: Dawsons. Published posthumously in 1727. Intro by I. Bernard Cohen.

151. Newton, Isaac. 1999. *The principia*. Trans. I. Bernard Cohen and Anne Whitman. Berkeley: University of California Press. This is a translation of the third edition of 1726, from which, using the footnotes, one is able to reconstruct the first (1687) and second (1713) editions.

152. Nordmann, Charles. 1922. Einstein: expose et Discute sa Théorie. *La Revue des Dux Mondes* 9: 129–166.

153. Nye, Mary Jo (ed.). 1984. *The question of the atom: from the Karlsruhe congress to the first solvay conference, 1860–1911*. New York: American Institute of Physics. This is a compilation of primary sources selected and introduced by Nye. Ostwald's essay "Emancipation from Scientific Materialism" is introduced and reprinted on 335–354.

154. Nye, Mary Jo (ed.). 2003. *The Cambridge history of science. Volume 5: the modern physical and mathematical sciences*. Cambridge: Cambridge University Press, 2003. Darrigol's essay is Chapter 17: Quantum Theory and Atomic Structure, 1900–1927. See Darrigol, 2003.

155. Ohanian, Hans C. 2009. Einstein's $E=mc^2$ Mistakes. This article was downloaded 27 January 2011 at: arxiv.org/ftp/arxiv/papers/0805/0805.1400.pdf. It has also been published as "Did Einstein prove $E=mc^2$?" *Studies in History and Philosophy of Science, Part B: Studies in History and Philosophy of Modern Physics*, vol. 40, 167–173.

156. Ogawa, Tsuyoshi. 1979. Japanese evidence for Einstein's knowledge of the Michelson-Morley experiment. *Japanese Studies in the History of Science* 18: 73–81.

157. Okun, Lev. B. 1989. The concept of mass. *Physics Today* (June), 31–36.

158. Okun, Lev. B. 2008. The Einstein formula: $E_0 = mc^2$: "Isn't the Lord Laughing?". This article was published in Russian and English in Russian journals in 2008. I downloaded an English version on 4 August 2008 at: arXiv:0808.0437v1 [physics.hist-ph].

159. Oppenheimer, J. Robert. 1965. On Albert Einstein. In A.P. French (ed.), *Einstein: a centenary volume*. Cambridge: Harvard University Press, 1979, pp. 44–49.

160. Oppenheimer, J. Robert. 1980. *Robert oppenheimer: letters and recollections,* ed. A.K.Smith, and C. Weiner. Cambridge, MA: Harvard University Press.

161. Ostwald, Wilhelm. 1895. Emancipation from scientific materialism. Reprinted in Nye (ed.), 1984, 335–354. The essay was delivered in 1895, and published in *Science Progress*, vol. IV, no. 24 (February, 1896). Trans. F.G. Donnan and F.B. Kenrick.

162. Pais, Abraham. 1982. *"Subtle is the Lord...": the science and the life of Albert Einstein*. Oxford: Oxford University Press.

163. Pais, Abraham. 1991. *Niels Bohr's times: in physics, philosophy, and polity*. Oxford: Oxford University Press.

164. Parker, Barry. 2003. *Einstein: the passions of a scientist*. New York: Prometheus Books.

165. Payne-Gaposchkin, Cecilia. 1996. *Cecilia Payne-Gaposchkin: an autobiography and other recollections,* ed. Katherine Haramundanis, 2nd ed. Cambridge: Cambridge University Press. The first edition was published in 1984.
166. Petersen, Aage. 1963. The philosophy of Niels Bohr. *The Bulletin of the Atomic Scientists* 19(7): 8–14.
167. Poincaré, Henri. 1905. The principles of mathematical physics. *The Monist* 15(1): 1–24 (January). This is an English translation of the 1904 lecture delivered in St. Louis.
168. Poincaré, Henri. 1952. *Science and hypothesis.* New York: Dover. This is a reprint of the 1905 English translation of the French edition of 1902.
169. Pyenson, Lewis. 1985. *The young Einstein: the advent of relativity.* Bristol/Boston: Adam Hilger.
170. Realdi, Matteo, and Giulio Peruzzi. 2008. Einstein, de Sitter and the beginning of relativistic cosmology in 1917. General relativity and gravitation. doi:10.1007/s10714-008-0664-y. Pdf file sent from the author in August 2010.
171. Reiser, Anton. 1930. *Albert Einstein: a biographical portrait.* New York: Albert & Charles Boni. The author's name is a pseudonym for Rudolf Kayser, Einstein's son-in-law, married to Elsa's daughter, Ilse. See Kayser, 1946.
172. Renn, Jürgen (ed.). 2005. *Albert Einstein, chief engineer of the universe. Volume I. One hundred authors for Einstein; vol. II. Einstein's life and work in context; vol. III. Documents of a life's pathway.* Weinheim: Wiley-Vch Verlag.
173. Rigden, John S. 2005. *Einstein 1905: the standard of greatness.* Cambridge, MA: Harvard University Press.
174. Rosenkranz, Ze'ev. 2005. Albert Einstein and the German Zionist Movement. In Renn (ed.), 2005, above, vol. I, 302–307.
175. Rothman, Tony. 2003. *Everything's relative: and other fables from science and technology.* Hoboken: Wiley.
176. Rubin, Vera C. 1980. Stars, galaxies, cosmos: the past decade, the next decade. *Science* 209(4452): 63–71 (Centennial Issue: July 4).
177. Sandage, Allan. 2004. *Centennial history of the Carnegie Institution of Washington: vol. I, The Mount Wilson observatory.* Cambridge: Cambridge University Press.
178. Sayen, Jamie. 1985. *Einstein in America: the scientist's conscience in the age of Hitler and Hiroshima.* New York: Crown Publishers.
179. Schilpp, Paul Arthur (ed.). 1949. *Einstein: philosopher-scientist.* Two volumes. New York: Harper & Row. I have used the 1959 Harper Torchbook edition. This work is a collection of essays on Einstein's legacy written mainly by noted scientists and philosophers. Several essays are cited in this book: those by Bohr, Born, Lemaître, Sommerfeld, and others. The first essay is the original version of Einstein's autobiography that Schilpp cajoled Einstein to write. I have used the corrected version (cited above, see Einstein, 1979) for most translations in English. Both it and the first volume of Schilpp also contain the original German version of the autobiography, which I have used when I did not agree with the published English translation.
180. Schweber, Silvan S. 2008. *Einstein and Oppenheimer: the meaning of genius.* Cambridge, MA: Harvard University Press.
181. Segre, Michael. 1998. The never-ending Galileo story. In *The Cambridge companion to Galileo,* ed. Peter Machamer, 388–416. Cambridge: Cambridge University Press.
182. Shankland, R.S. 1963. Conversations with Albert Einstein. *American Journal of Physics* 31: 47–57.
183. Sharlin, Harold I. 1966. *The convergent century: the unification of science in the nineteenth century.* London: Abelard-Schuman. This is a marvelous, albeit old-fashioned, account of the subject. It is out of print, but fortunately often available, dirt-cheap, on Amazon.
184. Shelton, Jim. 1988. The role of observation and simplicity in Einstein's epistemology. *Studies in History and Philosophy of Science* 19(1): 103–118.
185. Smith, Robert W. 1990. Edwin P. Hubble and the transformation of cosmology. *Physics Today* 43: 52–58 (April).

186. Smith, Robert W. 2003. See Kragh and Smith.
187. Smolin, Lee. 2004. Einstein's lonely path. *Discovery* 25(9): 36–41 (September).
188. Solovine, Maurice. 1986. *Letters to Solovine*. See Einstein, 1986, above.
189. Sommerfeld, Arnold. 1949. To Albert Einstein's seventieth birthday. In Schilpp (ed.), 1949, above, vol. 1, 99–105.
190. Stachel, John. 1987. Einstein and aether drift experiments. *Physics Today* 40(May): 45–47.
191. Stachel, John (ed.). 1998. *Einstein's miraculous year: five papers that changed the face of physics*. Princeton: Princeton University Press. I use, and cite, this book for the most recent translations of Einstein's 1905 papers.
192. Stachel, John. 2002. *Einstein from 'B' to 'Z'*. Boston/Basel/Berlin: Birkhäuser. This is volume Nine in the Einstein Studies series.
193. Staley, Richard. 2008. *Einstein's generation: the origins of the relativity revolution*. Chicago: University of Chicago Press. This important book sets Einstein's work within the context of the cultural, social, intellectual, material, theoretical, and experimental framework of his times, purposefully tempering the great man approach to science history. In light of – or, perhaps, in spite of – Staley's thesis, the title of this book was deliberately and carefully wrought.
194. Sugimoto, Kenji. 1989. *Albert Einstein: a photographic biography*. Translated from the German by Barbara Harshav. New York: Schocken Books.
195. Taylor, Edwin F., and John Archibald Wheeler. 1966. *Spacetime physics*. San Francisco/London: W. H. Freeman & Co.
196. Topper, David R. 1988. Review of T.S. Kuhn, *Black-body theory and the quantum discontinuity, 1894–1912* (Chicago, 1987 reprint). *Annals of Science* 45: 547–548.
197. Topper, David R. 2002. (History of) physics. *History of modern science and mathematics* 1: 115–146. Edited by Brian S. Baigrie. New York: Charles Scribner's Sons.
198. Topper, David R. 2007. *Quirky sides of scientists: true tales of ingenuity and error in physics and astronomy*. New York: Springer.
199. Topper, David R., and Dwight Vincent. 2000. Posing Einstein's question: questioning Einstein's pose. *The Physics Teacher* 38: 278–288 (May).
200. Topper, David R, and Dwight Vincent. 2007. Einstein's 1934 two-blackboard derivation of energy-mass equivalence. *American Journal of Physics* 75(11): 978–983 (November).
201. Trimble, Virginia. 1990. History of dark matter in the universe (1922–1974). In *Modern cosmology in retrospect,* ed. B. Bertotti and R. Balbinot, et al. Cambridge: Cambridge University Press, 355–362. Despite the title, she traces the idea back into the eighteenth century.
202. Van den Bergh, Sidney. 2011. The curious case of Lemaître's equation no. 24, submitted on 6 Jun 2011 to the *Journal of the Royal Astronomical Society of Canada*. Downloaded on 11 June 2011 at: http://arxiv.org/abs/1106.1195.
203. Van Dongen, Jeroen. 2007. Reactionaries and Einstein's fame: 'German scientists for the preservation of pure science', relativity, and the Bad Nauheim meeting. *Physics in Perspective* 9: 212–230.
204. Van Dongen, Jeroen. 2009. On the role of the Michelson–Morley experiment: Einstein in Chicago. *Archive for the History of the Exact Sciences* 63: 655–663. The copy I used was published online 17 July 2009, 9 pages.
205. Van Dongen, Jeroen. 2010. *Einstein's unification*. Cambridge: Cambridge University Press.
206. Walter, Scott. 1999. Minkowski, mathematicians, and the mathematical theory of relativity. In *The expanding worlds of general relativity*, ed. H. Goenner and J. Renn, et al. Boston/Basel/Berlin: Birkhäuser. This is volume Seven in the Einstein Studies series.
207. Weinberg, Stephen. 2011. The universes we still don't know, essay review of *The grand design*, by Stephen Hawking and Leonard Mlodinow (Bantam Books), *The New York Review of Books*, vol. LVIII, no. 2 (February 10), 31–34.
208. Whittaker, Edmund. 1955. Albert Einstein, 1879–1955. *Biographical Memoirs of Fellows of the Royal Society* 37–67.
209. Whittaker, Edmund. 1960a. *A history of the theories of aether and electricity, volume I: the classical theories*. New York: Harper Torchbook. This was originally published in 1910, and revised and enlarged in 1951.

210. Whittaker, Edmund. 1960b. *A history of the theories of aether and electricity, volume II: the modern theories, 1900–1926*. New York: Harper Torchbook. This was originally published in 1953. Relativity is covered in Chapter II, with the title: "The Relativity Theory of Poincaré and Lorentz," which betrays Whittaker's anti-Einstein bias.

211. Will, Clifford M. 1990. General relativity at 75: how right was Einstein? *Science* 250: 770–776 (November).

212. Will, Clifford M. 1993. *Was Einstein right?: putting general relativity to the test*, 2nd ed. New York: Basic Books.

213. Williams, L. Pearce. 1971. *Michael Faraday: a biography*. New York: A Clarion Book.

214. Winteler-Einstein, Maja. 1987. Albert Einstein – A biographical sketch (Excerpt). *Einstein Papers* 1: xv–xxii. See Einstein, 1987, above. Einstein's sister, Maja, wrote this sketch in 1924.

215. Woolf, Harry (ed.). 1980. *Some strangeness in the proportion: a centennial symposium to celebrate the achievement of Albert Einstein*. Reading: Addison-Wesley. See especially Section XI: Working with Einstein, 473–489.

216. Wright, Helen. 1970–80. Adams, Walter Sydney. In *Dictionary of scientific biography*, ed. C.C. Gillispie. New York: Charles Scribner's Sons, vol. I, 54–58.

217. Yam, Philip. 2004. Everyday Einstein. *Scientific American* 291(3): 50–55 (September).

218. Young, Thomas. 1807. *A course of lectures on natural philosophy and the mechanical arts*. Two volumes. London: for Joseph Johnson, St. Paul's Church Yard, by William Savage. Unfortunately, the text does not give the dates when each lecture was delivered. I used the Johnson Reprint Publication of 1971, which is No. 82, in the Sources of Science series.

219. Zackheim, Michele. 1999. *Einstein's daughter: the search for Lieserl*. New York: Riverhead Books.

220. Zwicky, F. 1933. Die Rotverschiebung von extragalaktischen Nebein. *Helvetica Physica Acta* 6: 110–127.

Index

A

Aarau, 11, 12, 15, 212
AAS. *See* American Astronomical Society
 (AAS)
Abiko, Seiya, 52, 69, 89
Absolute motion, 9, 45, 97, 98, 100
Absolute rest, 44–47
Absolute space, 69, 97–101, 175, 199, 200,
 209, 211
Absolute time, 28, 56, 69
Académie des Sciences, 129
Action-at-a-distance, 25, 26, 28, 30, 38, 40,
 83, 195, 196, 199, 203, 229
Adams, John C., 110
Adams, Walter S., 140
Adler, Carl G., 80
Advance of the perihelion, 110, 115, 117
Aether, 13, 23, 28, 30, 35–38, 46, 47,
 49–52, 55, 59, 60, 63, 64, 68,
 80, 85, 123, 140, 157, 195,
 196, 200–203, 209
Aether-drift experiment, 140
Almeida, José B., 189
Alpher, Ralph, 165, 183
American Astronomical Society (AAS), 157,
 158, 167
Ampère, André-Marie, 31
Andromeda nebula, 150, 156, 158, 171
Annalen Der Physik (Annals of Physics), 19
Annus Mirabilis, 15–23, 41
Anti-relativity movement, 124
Anti-semitism, 75, 122, 131
Aristarchus, 3
Aristotle, 26, 85–87, 149
Arnheim, Rudolf, 95
Aryan science, 123
Astrology, 26, 29

Astronomical unit (AU), 152, 153
Atom, 19, 21, 41, 62, 73, 94, 175, 176, 180,
 209–211, 218, 220, 221, 232
Atomic, 21, 62, 179, 211, 221, 227
Atomic clock, 79, 94, 118, 143, 144
Atomism. *See* Atom
AU. *See* Astronomical unit (AU)

B

Barbour, Julian, 97
Bargmann, Valentin (Valya), 218
Barnett, Lincoln, 80
Bartusiak, Marcia, 151, 156–158, 166, 172
Bell Labs, 183
Bergmann, Peter, 218
Bergson-Einstein debate. *See* Einstein-Bergson
 debate
Bergson, Henri, 127
Berkeley, George, 228
Berlin, 75, 76, 108, 121–124, 140, 147, 191,
 210, 215, 217, 225, 231
Bern, 15–23, 67, 69, 73, 76, 89, 108
Bernstein, Jeremy, 184
Besso, Michele Angelo, 15, 18, 60, 69, 73, 97,
 116, 230
The Bible, 10
Big bang model, 181–183, 186–189
Black-body radiation, 20, 51, 64
Black hole, 184, 186
Blueshift, 156, 157
Bohr, Niels, 133, 206, 221–223, 225–230
Bohr-Einstein debate, 221, 228, 229
Bologna, 77
Bólyai, Janos, 105
Bondi, Hermann, 97, 180, 183
Born, Hedwig, 75

D.R. Topper, *How Einstein Created Relativity out of Physics and Astronomy*, Astrophysics
and Space Science Library 394, DOI 10.1007/978-1-4614-4782-5,
© Springer Science+Business Media New York 2013

Born, Max, 40, 75, 223, 226, 228, 230
Bose, Satyendra Nath, 221
Bose-Einstein statistics, 221
Brahe, Tycho, 103
Brazil, 121
Brian, Denis, 215
Brownian motion, 19, 20, 133
Brush, Stephen, 20, 125, 132, 133, 181, 221–223, 228, 229
Bucherer, Alfred, 65
Bucky, Peter A., 147, 217, 233

C

Calculus, 11, 35, 41, 103, 107, 110, 115, 119, 121, 159, 160, 165, 168, 201, 204, 210
Calder, Nigel, 146
Caloric, 37
Caltech, 137–141, 171–177, 191, 199–206, 215
Cambridge University, 41, 121, 180
Campbell, William W., 140
Canales, Jimena, 131, 134
Carnegie Institute of Technology, 197
Carnegie-Mellon University, 197
Carter, Paul, 216, 219
Case-Western Reserve University, 49
Cassidy, David C., 109, 123, 135
Catholic church, 74, 208
Cavendish, Henry, 29, 92, 195
Centrifugal force, 98, 101, 102, 145, 188
Cepheid variables, 154–156, 158, 171, 173
Cephius. *See* Cepheid variables
CERN, 145, 146
Cervantes, 74, 233
Chaplin, Charlie, 138
Chicago, 53, 125, 132, 137, 140, 157
Clark, Ronald W., 70, 76, 90, 217, 230
Clock paradox, 77, 78, 128
COBE. *See* Cosmic Background Explorer (COBE) satellite
Cohen, I. Bernard, 217
Collège de France, 77, 129
Compass, 26, 31, 32, 40, 95, 100
Complementarity principle, 225
Comte, Auguste, 208
Conservapedia, 125
Conservation of energy, 38, 40, 61
Copernicus, Nicholas, 3, 14, 27, 149, 207
Cosmic Background Explorer (COBE) satellite, 185, 186, 189
Cosmic microwave background radiation, 183
Cosmic rays, 79, 145, 175, 176
Cosmological constant, 161–166, 169, 174–176, 189, 190, 199

Coulomb, Charles-Augustin, 29
Coulomb's law, 29, 32, 35
Covariance, 7, 60, 62, 111, 212, 213. *See also* Invariance
Crelinsten, Jeffrey, 121, 122, 137, 140
Critical model, 187, 188
The Critique of Pure Reason (Kant), 11, 29, 212
Cuba, 138
Curved space, 204–206
61 Cygni, 152, 153

D

Dark energy, 188, 189
Dark matter, 188, 189
Darnton, Robert, 26
Darrigol, Olivier, 20
Darwin, Charles, 125
Darwin, George, 180
Dead of Night (movie), 180
De Brogile, Louis, 224
De Magnete (Gilbert), 26
De Sitter, Willem, 162, 166, 168, 172, 174, 175
Dewhirst, David, 157
Dicke, Robert, 183, 189
Dickens, Charles, 74
Don Quixote (Cervantes), 74, 233
Donley, Carol C., 122, 238
Doppler principle (or effect), 94, 157, 174, 188
Duerbeck, Hilmer W., 161
Dukas, Helen, 137, 193, 215, 216, 219, 234
Durée et Simultanéité: A Propos de Théorie d'Einstein (Bergson), 130
Dyson, Freeman, 68

E

Eddington, Arthur S., 121, 161, 166–168, 175, 217
Ehrenfest, Paul, 20, 65, 74, 109, 111, 124, 129, 147, 162, 229, 230
Eidgenössische Technische Hochschule (ETH), 15, 39, 51, 64, 70, 73, 76, 107, 147, 173
Einstein, Albert, *passim*,
Einstein archives, 51, 80, 137, 189, 201, 202
Einstein, Eduard (Tete or Tedel), 19, 216
Einstein, Elsa Löwenthal, 109, 117, 122, 124, 131, 137, 191, 192, 215, 216
Einstein, Hans Albert, 19
Einstein, Hermann, 9–11, 39, 68
Einstein, Ilse, 215, 216
Einstein, Lieserl, 17, 23, 51
Einstein, Margot, 215, 216
Einstein, Marie (Maja), 9, 11, 216

Einstein, Mileva Marić, 15–18, 23, 46, 51, 73, 76, 109, 117, 123
Einstein, Pauline, 9
Einstein-Bergson debate, 134
Einstein-Bohr debate. *See* Bohr-Einstein debate
Eisenstaedt, Jean, 184
Electric current, 32, 34, 35, 44
Electric field, 34, 43, 44, 89, 93
Electric forces, 33, 46
Electric generator (or dynamo), 34
Electricity, 25, 26, 29–32, 34–39, 43, 46, 62, 68, 91, 114, 135, 147, 190, 195, 196, 201–203, 206, 218, 220
Electric motor, 34, 89
Electromagnet, 31–34
Electromagnetic field, 34–36, 133, 134, 191, 195, 197, 201, 205
Electromagnetic waves, 36, 40
Electron, 21, 62, 63, 65, 69, 72, 73, 79, 81, 129, 218, 224, 228
Empty space, 13, 28, 30, 45, 46, 83, 90, 93, 102, 152, 166, 169, 198, 200, 203
Energeticism, 39, 72, 196
Energeticist, 39
Energy, 6, 19, 20, 36–40, 60–62, 64, 72, 79, 80, 116, 159, 181–184, 186–189, 195–197, 205, 219–221, 223
Eötvös, Baron Roland von, 92
Eötvös experiment, 92, 144, 145
Epistemology, 133, 207, 210, 222, 223, 229
Equivalence principle (or postulate), 94, 112, 114, 123, 143–145, 163, 200, 210. *See also* Principle of equivalence
ETH. *See* Eidgenössische Technische Hochschule (ETH)
Ethics (Spinoza), 74, 75, 217, 226
Euclid, 45, 95, 105, 211
Euclidean geometry, 11, 30
Euclidean space, 49, 149, 150, 169
Expanding model, 94, 179, 180, 183–185

F
Faraday, Michael, 32, 217
Farrell, John, 167, 176, 189, 216
FBI, 220
Feuer, Lewis S., 74, 75
Feynman Lectures, 134
Feynman, Richard, 134
Field theory, 191–193
First World War, 108, 129, 157
Fishbane, Paul, 124
Fölsing, Albrecht, 7, 69, 117, 131, 215, 228

Foucault, Leon, 98–100, 102
Fourth dimension, 70–72, 104, 106, 117, 160, 218
Fowles, Grant R., 50
Francis W. Parker School, 53
Frank, Philip, 75, 124
Free-Fall, 89, 91, 93
French, A. P., 61, 238
Friedman, Alan J., 122, 238
Friedmann, Aleksandr, 166–168, 172, 175

G
Galaxy, 152–154, 156–158, 171, 172, 182, 186, 188
Galilei, Galileo, 2–7, 12, 14, 26, 36, 41, 44, 45, 83–88, 90–92, 97, 99, 101, 102, 116, 127, 145, 149, 151, 208, 211, 217, 233
Galison, Peter, 56, 67–69, 72, 77, 193
Gamow, George, 165, 183
Gandhi, 217
Gauss, Carl Friedrich, 105
General relativity, 69, 78, 83–84, 92, 94, 104, 107, 108, 113–119, 127, 132, 133, 140, 143, 144, 159, 160, 162, 163, 165, 166, 169, 184, 187, 188, 196, 197, 199–202, 204, 210, 212, 213, 218
Geocentric cosmos (or model), 26, 100
German Reichstag, 131
German University of Prague, 73
Gibbs, Josiah Willard, 197
Gilbert, William, 26
Global positioning satellite system (GPS), 79, 144
God, 20, 28, 75, 97, 98, 111, 122, 131, 150, 161, 175–177, 200, 219, 226, 230, 233
Gold, Thomas, 180, 183
GPS. *See* Global positioning satellite system (GPS)
Grafton, Anthony, viii, 239
Graham, Loren, 134, 135
Gravitational bending of light, 115, 138
Gravitational drag of light, 173
Gravitational red shift, 94, 115–117, 140, 143, 182
Gravitational time dilation, 94, 115, 117–119, 143
Gravity, 6, 27–29, 31, 32, 34–36, 40, 41, 74, 75, 78–80, 83, 90–94, 100–104, 106–111, 114–118, 128, 135, 143–145, 149, 150, 160, 163, 164, 167, 173, 182–184, 186–188, 195–199, 201–206, 208, 218, 220, 231, 232

Green, Jim, 109, 220
Grossmann, Marcel, 15, 17, 19, 75, 107, 113, 218, 219
Grundmann, Siegfried, 129
Gunter, P. A. Y., 130, 131

H

Habicht, Conrad, 18–20, 69, 111, 229, 230
Hadamard, Jacques, 95
Hale, George Ellery, 138, 140
Halpern, Paul, 218
Handel, George Friedrich, 150
Harman, P. M., 33, 38
Harvard Observatory, 154, 158
Hawking, Stephen, 134, 232
Heat, 22, 35, 37, 38, 40, 208
Hebrew University, 53, 122, 137
Heidegger, Martin, 134
Heidelberg University, 15, 124
Heisenberg, Werner, 227
Heliocentric cosmos (or model), 3, 26, 207
Helmholtz, Hermann von, 37, 38
Hentschel, Klaus, 64, 122, 141
Herbert, Christopher, 135
Herbert Spencer Lecture, 211, 216
Herman, Robert, 165, 183
Herschel, Caroline, 150–152, 156
Herschel, William, 110, 150
Hertz, Heinrich, 36
Hetherington, Norriss S., 140
Highfield, Roger, 216, 217, 219
Hilbert, David, 112
Hitler, 124, 191
Hoefer, Carl, 97, 174
Hoffmann, Banesh, 18, 73, 217, 218, 234
Holton, Gerald, 12, 43, 51–53, 64, 72, 137, 211, 220, 232
"Holy geometry book," 11, 40, 95, 104, 206
Hoskin, Michael, 152, 157
Howard, Don, 213
Hoyle, Fred, 180, 181, 183, 184
Hubble, Edwin P., 140, 141, 157, 158, 164, 168, 169, 171–177, 179, 181, 233
Humason, Milton L., 140, 141, 171–174, 179
Hume, David, 217
Huygens, Christiaan, 98

I

Industrial revolution, 37, 208
Inertia, 3–7, 12, 14, 59, 60, 85, 87, 88, 91, 99, 101, 102, 104, 145, 159, 163, 175, 199, 200, 203

Inertial force, 90, 117
Inertial mass, 91, 92, 117
Inertial system, 7, 9, 12–14, 45, 55, 60, 71, 78, 83, 107, 113, 127, 128, 130, 196, 197, 203
Infeld, Leopold, 6, 34, 55, 56, 115, 192, 196, 218, 229, 233
Infrared light, 22
Institute for Advanced Study, 6, 191, 192, 216
Instrumentalism. See Phenomenalism
Interference of light, 21, 22, 50, 132, 133, 202
Invariance, 7, 13, 14, 45, 56, 60, 62, 72, 134, 135. See also Covariance
Inverse-square law, 27–29, 41, 150, 156, 164, 188, 195, 204
Isaacson, Walter, 9, 17, 18, 70, 112, 131, 132, 215–217, 226
Island universe, 153, 154, 156–158, 171

J

Jaffe, Georg, 226
Jewish physics. See Aryan science
Jewish science. See Aryan science
Johan Sebastian Bach, 232
Jones, Sheilla, 230
Jungnickel, Christa, 103

K

Kafka, Franz, 74
Kahn, Carla, 166, 240
Kahn, Franz, 166, 240
Kaiser Wilhelm Institute of Physics, 76
Kaluza-Klein theory, 218, 219
Kaluza, Theodor, 217, 218
Kant, Immanuel, 29, 30, 34, 38, 195, 196, 217
Kaufman, Bruria, 218, 219
Kaufmann, Walter, 62, 63, 72, 73
Kayser, Rudolph (aka Anton Reiser), 40, 215, 236, 240, 243
Kelvin, Lord. See Thomson, William
Kepler, Johannes, 4, 103, 149, 156, 208, 211, 233
Kerszberg, Pierre, 166, 175, 177
Klein, Felix, 65
Klein, Martin J., 20, 74, 173, 221, 222, 224, 230
Koyré, Alexander, 150, 241
Kragh, Helge L., 164, 167, 168, 175, 182, 183, 185
Kuhn, Thomas S., 20, 37
Kyoto University, 52

L

Lange, Ludwig, 7
Langevin, Paul, 77, 127, 129–131, 224
Le Verrier, Urbain, 110, 111
Leavitt, Henrietta, 154–156, 158
Leavitt's law, 155, 156, 158, 172
Leibniz, Gottfried, 98
Leiden, 124, 166, 197, 199–201
Leiden University, 74
Lemaître, Georges, 167–169, 174–177, 189, 216
Lenard, Philipp, 16, 17, 123, 124
Lerner, Eric J., 189
Levenson, Thomas, 117, 215
Lick Observatory, 140, 156
Lieserl. *See* Einstein, Lieserl
Light-clock, 81
Light-year, 153, 155, 156, 158, 184
Livio, Mario, 168
Lobachevski, Nikolai I., 105
Local time, 27, 28, 56, 68, 209
Lodestone, 25, 26
London Times, 212
Lorentz, Hendrik A., 80, 200
Lorentz transformation, 80
Los Alamos, 220
Lowell Observatory, 154, 156, 168, 171
Lowell, Perceval, 156
Löwenthal, Elsa. *See* Einstein, Elsa

M

Mach, Ernst, 7, 45, 74, 94, 97–102, 113, 175, 199, 203, 207–213
Mach's principle, 6, 97–102, 113, 123, 163, 164, 166, 174, 175, 199, 203, 204, 209–211, 218
Magellanic clouds, 154
Magnetic field, 34
Magnetic forces, 29, 31–33, 198
Magnetism, 25, 26, 29–34, 37, 39, 40, 60, 147, 195, 196, 202, 203, 206
Marić, Mileva. *See* Einstein, Mileva Marić
Mars, 103
Martinez, Alberto A., 23
Mass, 6, 19, 27, 55, 79, 91, 101, 104, 113, 123, 128, 144, 150, 161, 167, 173, 184, 196, 200, 208, 224
Mass-energy, 61, 62, 71, 195
Matter and Memory (Bergson), 130
Matter-waves, 224
Maxwell, James Clerk, 35, 36, 38, 41, 195, 202, 217
Maxwell's equations, 35, 60, 62, 71, 157, 174, 195, 197, 198, 203, 205

Mayer, Walther, 137, 213, 215, 216, 218
McCormmach, Russell, 103, 233
McFarland, Ernie, 80
McVittie, George C., 168, 176, 189
The Meaning of Relativity (Einstein), 129, 219
Mechanical philosophy, 28, 59
Mechanical work, 37, 38
Mercury, 110, 111, 115–117, 143
Mesmer, Franz, 26
Metaphysical Foundations of Natural Science (Kant), 30
Michelmore, Peter, 123, 175, 216, 219
Michelson, Albert A., 49–52, 81, 125, 140, 199
Michelson-Morley experiment, 49, 50, 59, 63, 69, 140, 143
Microwave energy, 183
Milky Way, 151–153, 156–158, 163, 164, 171, 172, 182, 185
Miller, Arthur I., 11, 52, 68, 78, 95, 96, 130
Millikan, Robert, 132, 133, 137, 138, 176, 191, 223
Minkowski, Hermann, 70, 72, 73, 78, 107
Mlodinow, Leonard, 232
Moon, 3, 26–28, 30, 100, 108, 118, 151, 171, 207
Moszkowski, Alexander, 40, 41, 69
Mt. Everest, 118
Mt. Hamilton. *See* Lick Observatory
Mt. Palomar, 154, 180, 184
Mt. Wilson Observatory, 138, 155, 158
Munich, 9, 10
Muon, 79, 145, 146, 219

N

Nathan, Otto, 109, 137, 219
Nazi, 123, 124, 135, 191, 192, 215, 217, 234
Nebula (Nebulae), 150–154, 156–158, 166, 167, 169, 171–174, 180
Neffe, Jürgen, 9, 17, 19, 75, 76, 192, 193, 215, 216
Neptune, 110, 153
Neutrino, 219
Neutron, 219
Neutron star. *See* Pulsars
Newtonian physics, 28, 144, 231
Newton, Isaac, 6, 21, 27–29, 35, 36, 41, 55, 59, 61, 72, 83, 91–93, 97–99, 101, 102, 105, 110, 116, 118, 122, 145, 149–158, 161, 163, 169, 195, 196, 208, 210, 217, 233, 234
Newton's bucket experiment, 97, 98, 102
New York City, 138

The New York Times, 90, 141, 173, 206, 226
Nobel Prize, 16, 63, 69, 123, 125, 132, 183,
    192, 221, 232
Non-Euclidean geometry, 105, 106, 117, 119,
    201, 205, 206, 218
Non-Euclidean space, 104–107, 115, 128, 169
Noninertial system, 90, 91, 93, 101, 104, 114,
    117, 128, 163
Norden, Heinz, 109
Nordmann, Charles, 129, 131
North star. *See* Polaris
Nova. *See* Supernova
Nuclear (strong) force, 220, 232
Nuclear physics, 21, 179, 180, 220, 232
Nucleus, 21, 180, 220
Nye, Mary Jo, 39

O
Occult forces, 30, 83, 93, 200
Øersted, Hans Christian, 30–32, 34, 35, 38
Ogawa, Tsuyoshi, 52
Ohanian, Hans S., 80, 125
Okun, Lev B., 80
Olympia Academy, 18, 68, 74, 130, 233
Ono, Yoshimasa A., 52
Ontology, 222, 223, 227, 229
Oppenheim, Paul, 122
Oppenheimer, Robert J., 96, 184, 218
Orion, 151
Ostwald, Wilhelm, 39, 72, 195, 196
Oxford University, 231

P
Padua, 85
Pais, Abraham, 19, 53, 125
Parallax. *See* Stellar parallax
Paris Observatory, 110, 129
Parker, Barry, 215
Particle model, 21, 22, 35, 133, 137, 220
Pasadena, 137, 138, 140, 206
Patent office, 17, 18, 56, 67–69, 73, 89, 96, 209
Payne-Gaposchkin, Cecelia, 158
Peebles, Jim, 183, 189
Pendulum, 59, 81, 91, 98–101, 145
Penzias, Arno, 183
Perihelion, 10, 111, 115–117, 143
Period-luminosity law, 155
Peruzzi, Giulio, 166
Petersen, Aage, 229
Pfister, Herbert, 97
Phenomenalism, 207–209, 211, 213, 223, 225,
    227–229. *See also* Positivism

Photoelectric effect, 17, 132, 133, 157
Photons, 65, 81, 133, 176, 218, 223, 224
Pierpont Morgan library, 44, 83
Pisa, 3, 85–87, 145
Planck, Max, 20, 65, 69, 124, 210
Plato, 127
Podolsky, Boris, 229
Poincaré, Henri, 68, 69, 74, 125
Polaris, 26, 100
Popper, Karl, 134
Positivism, 207–213, 223, 225, 228, 229, 232
Positivist, 208, 209, 212, 227–229
Prague, 73–75, 107, 108, 122, 226
Princeton University, 129, 182, 191, 219
Principe Island, 121
Principia, 6, 27, 28, 30, 36, 45, 97, 150, 208, 234
Principle of equivalence, 112, 114, 117, 123,
    143–145, 163, 200, 210
Principle of relativity, 9, 13, 14, 44–47, 50, 51,
    55, 63, 65, 68, 78, 91–93, 101, 163, 197
Prussian Academy of Science, 76, 109, 113
Ptolemy, 149
Pulsars, 184
Pyenson, Lewis, 11, 12, 73

Q
Quantum mechanics, 226–228, 231–233
Quantum of energy, 20, 221, 223
Quantum of light, 132, 224
Quasar, 143, 182–184, 186

R
Radcliffe College, 154
Radiant heat, 22, 37
Radio waves, 22, 181, 182
Rathenau, Walther, 129, 131
Realdi, Matteo, 166
Realism, 153, 207–209, 211, 213, 225,
    227–229, 232
Redshift, 94, 115–118, 140, 141, 143, 157,
    166, 169, 172–174, 179, 180–184,
    186, 188
Reiser, Anton. *See* Kayser, Rudolph
Relativism, 45, 128, 134, 135
Relativity principle. *See* Principle of relativity
Renn, Jürgen, 10, 123, 147, 217, 218
Riemann, Bernhard, 105, 201
Rosenkranz, Ze'ev, 122
Rosen, Nathan, 229
Rothman, Tony, 22, 62, 110, 122
Royal Institution, 22
Royal Society of London, 121

Rubin, Vera, 184
Russell, Bertrand, 134
Rutherford, Ernest, 21

S
Samuel, Joseph, 97
Sandage, Allan, 171, 172, 181, 206
San Jose, California, 140
Sartre, Jean-Paul, 134
Saturn, 28
Sayen, Jamie, 193, 233
Scandinavian Society of Science, 133
Schilpp, Paul Arthur, 10, 134, 189, 201, 223, 225–228
Schrödinger, Erwin, 227
Schubert's unfinished symphony, 232, 233
Schweber, Silvan S., 96, 135, 220
Science (Journal), 184, 206
Science and Hypothesis (Poincaré), 68
The Science of Mechanics (Mach), 97
Science Wars, 134
Scientific American, 165, 205, 219
Second World War, 118, 124, 138, 192, 226
Seelig, Carl, 201, 237
Segre, Michael, 86
Seitter, Waltraut C., 161
Shankland, Robert S., 217
Shapley, Harlow, 155–158, 167, 171
Shelton, Jim, 46
Simultaneity, 56, 67, 130, 131
Sinclair, Upton, 138
Sirius B, 140, 144
Slipher, Vesto M., 156, 157, 166, 168, 169, 171, 172, 181
Smith, Robert W., 167, 168, 181
Smolin, Lee, 233
Society of French Physicists, 129
Sokal, Alan, 134
Solar eclipse, 107, 108, 117, 121, 138, 140
Solar system, 110, 111, 149, 151, 153, 154, 156
Solovine, Maurice, 17, 18, 68, 74, 122, 130, 131, 216, 232, 233
Sommerfeld, Arnold, 51, 73, 107, 134, 135, 226
Sorbonne, 130
Space, 4, 9, 18, 28, 45, 49, 55, 68, 77, 83, 90, 97, 104, 113, 128, 141, 144, 149, 160, 165, 171, 180, 195, 199, 208
Space-time, 67–76, 104, 106, 115, 117, 119, 176, 195, 199, 206, 210
Special relativity, 44–46, 52, 57, 58, 60, 69, 79, 83, 89, 93, 107, 113, 116–119, 125, 127, 143, 145, 157, 162, 168, 169, 197, 200, 203, 210, 221

Spectroscope, 153, 154, 156, 159, 207
Spectrum of light, 21, 94, 117, 153, 188
Speed of light, 1, 12–14, 36, 45, 56–63, 65, 68, 71, 72, 77, 79, 94, 107, 118, 130, 140, 146, 162, 182, 196, 231
Spinoza, Baruch, 74, 75, 217, 226
Stachel, John, 17, 40, 43, 44, 51, 57, 58, 60, 61, 64, 65, 69, 74, 83, 97, 107, 205, 219, 221, 223, 224, 226
Staley, Richard, 51, 69, 73, 81
Star clusters, 151, 156
Stark, Johannes, 63, 64, 69, 89, 93, 124
Steady state model, 180–184, 186
Steam engine, 37
Stellar parallax, 98, 100, 152, 153
St. John, Charles E., 140, 141
St. Louis, 68
Stonehenge, 100
Strauss, Ernst G., 218
The Structure of Scientific Revolutions (Kuhn), 20
Subatomic particles, 79, 145, 176, 229, 232
Sugimoto, Kenji, 217
Sun, 3, 21, 22, 26–28, 30, 65, 79, 94, 99, 107, 108, 110, 111, 115–118, 121, 138, 141, 143, 144, 149, 152, 153, 171, 176, 180, 184, 197, 207, 208, 215
Supernova, 125, 184, 188
Swiss Polytechnic Institute. *See* Eidgenössische Technische Hochschule (ETH)

T
Taylor, Edwin F., 28, 244
Thermodynamics, 37, 38, 147, 208, 222
Thomson, G. P., 224
Thomson, J. J., 62, 224
Thomson, William, 38
Thought-experiment, 1, 4, 9–14, 39, 40, 44, 45, 51, 52, 56, 83, 85–87, 89–96, 97, 98, 100, 102–104, 113, 114, 119, 128, 144, 145, 191, 200, 206
Tides, 26, 29, 208
Time, 3, 9, 15, 25, 45, 49, 55, 67–79, 85, 90, 98, 103, 113, 121, 127–135, 138, 143, 150, 159, 165, 171, 179, 195, 200, 208, 215, 233
Time and Free Will (Bergson), 130
Time dilation, 57–60, 62, 77–79, 81, 94, 115, 117–119, 125, 127, 129, 130, 143, 145, 146
Tired-light hypothesis. *See* Gravitational drag of light

Topper, David R., 12, 20–21, 25, 63, 116, 118,
    122, 151, 156–158, 166-168,173–174,
    181–183, 185, 192, 197, 204, 206, 219,
    232, 234, 244
Trimble, Virginia, 188
Twin paradox. *See* Clock paradox

U
Ulm, 9
Ultraviolet light, 17, 22, 64
Unified field theory, 35, 107, 119, 133, 175,
    192, 201, 203–206, 220, 231–233
University of Glasgow, 83, 216
University of Vienna, 20
Uranus, 110, 150

V
Van den Bergh, Sidney, 168
Van Dongen, Jeroen, 53, 64, 107, 213, 218,
    219, 223
Vectors, 35, 205
Victorian Relativity (Herbert), 135
Vincent, Dwight, 174, 181, 197, 204–206
Von Laue, Max, 69, 78
Von Neumann, John, 192

W
Walter, Scott, 73
Warped space. *See* Curved space
Wave model, 21–23, 116, 223
Weak (nuclear) force, 232
Weight, 3, 4, 27, 28, 59, 62, 71, 85–92,
    95, 116

Weinberg, Steven, 232
Wheeler, John A., 28, 183
Whitehead, Alfred North, 134
Whittaker, Edmund, 68, 69, 201
Wien, Wilhelm, 51
Wikipedia, 125
Wilkinson, David, 183, 189
Wilkinson Microwave Anisotropy Probe
    (WMAP), 189
Will, Clifford, 80, 125, 143, 144, 146
Williams, L. Pierce, 34, 35
Wilson, Robert W., 183
Winnipeg, 100, 152
Winteler-Einstein, Maja. *See* Einstein, Marie
WMAP. *See* Wilkinson Microwave Anisotropy
    Probe (WMAP)
Woolf, Harry, 218, 234
Wright, Helen, 140

X
X-rays, 22, 69, 186

Y
Yam, Philip, 144
Yerkes Observatory, 157
Young, Thomas, 21, 22, 36

Z
Zackheim, Michele, 17
Zionism, 53, 74, 75, 122
Zürich, 11, 15–23, 51, 70, 73, 75–77, 108,
    124, 127
Zwicky, Fritz, 173, 181, 188